AUGUSTUS DE MORGAN

THE CONNEXION OF NUMBER AND MAGNITUDE

AN ATTEMPT TO EXPLAIN THE FIFTH BOOK OF EUCLID

Elibron Classics
www.elibron.com

Elibron Classics series.

ISBN 1-4021-8696-7 (paperback)
ISBN 1-4021-0796-X (hardcover)

This Elibron Classics Replica Edition is an unabridged facsimile of the edition published in 1836 by Taylor and Walton, London.

CONNEXION

OF

NUMBER AND MAGNITUDE:

AN ATTEMPT TO EXPLAIN

THE FIFTH BOOK OF EUCLID.

BY

AUGUSTUS DE MORGAN,

OF TRINITY COLLEGE, CAMBRIDGE.

La seule manière de bien traiter les élémens d'une science exacte et rigoureuse, c'est d'y mettre toute la rigueur et l'exactitude possible. — D'ALEMBERT.

LONDON:

PRINTED FOR TAYLOR AND WALTON,

BOOKSELLERS AND PUBLISHERS TO THE UNIVERSITY OF LONDON,

30 UPPER GOWER STREET.

M.DCCC.XXXVI.

LONDON:
J. MOYES, CASTLE STREET, LEICESTER SQUARE.

PREFACE.

THIS Treatise is intended ultimately to form part of one on Trigonometry. The place in which most students consider number and magnitude together for the first time, is in the elements of the latter science, unless they have understood the Fifth Book of Euclid better than is usually the case. Previously, therefore, to commencing Trigonometry, I consider it advisable to enter upon the consideration of proportion in its strict form; that is, upon the Fifth Book of Euclid. There is no other method with which I am acquainted which gives any thing like demonstration of the general properties of ratios, though there is a *doux oreiller pour reposer une tête bien faite*, which many of the continental mathematicians have agreed shall be called demonstration, and which is beginning to make its way in this country.

Hitherto, however, it has been customary for mathematical students among us to read the Fifth Book of Euclid; frequently without understanding it. The form in which it appears in Simson's edition is certainly unnecessarily long, and the tedious repetition of "AB is the same multiple of CD which EF is of GH," in all the length of words, renders the reasoning not easy to follow. The use of general symbols of concrete magnitude, instead of the straight line of Euclid, and of a general algebraical symbol for whole number, seems to me to remove a great part of the difficulty. Throughout this work it must be understood, that a capital letter denotes a magnitude; not a numerical representation, but the magnitude itself: while a small letter denotes a number, and mostly a whole number. And by the term *arithmetical proportion*, when it occurs, is signified, not the common and now useless meaning of the words, but the proportion of two magnitudes which are arithmetically related, or which are commensurable.

The subject is one of some real difficulty, arising from the limited character of the symbols of arithmetic, considered as representatives of ratios, and the consequent introduction of incommensurable ratios, that is, of ratios which have no arithmetical representation. The whole number of students is divided into two classes: those who do not feel satisfied without rigorous definition and deduction; and those who would rather miss both that take a long road, while a shorter one can be cut at no greater expense, than that of declaring that there *shall be* propositions which arithmetical demonstration declares there *are not*. This work is intended for the former class.

AUGUSTUS DE MORGAN.

London, May 1, 1836.

ERRATA.

Page 35, line 3, *for* less *read* greater.
...... 47, ... 1, *for* $v(Q + Z)$ *read* $w(Q + Z)$.
...... — ... 2, *for* vQ *read* wQ.

CONNEXION

OF

NUMBER AND MAGNITUDE.

WHEN a student has acquired a moderate knowledge of the operations and principles of algebra, with as many theorems of geometry as are contained in the first four books of *Euclid's Elements*, it becomes most desirable that he should gain some more exact knowledge of the connexion between the ideas which are the foundation of one and the other science, than would present itself either to an inattentive reader, or to one whose whole attention is engrossed by the difficulty of comprehending terms which cannot yet have become familiar to him. Before proceeding, therefore, to explain Trigonometry (the measurement of triangles), which, in the widest sense, includes all the applications of algebra to geometry, it will be right to inquire on what sort of demonstration we are to pass from an arithmetical to a geometrical proposition, or *vice versâ*.

Geometry cannot proceed very far without arithmetic, and the connexion was first made by Euclid in his Fifth Book, which is so difficult a speculation, that it is either omitted, or not understood by those who read it for the first time. And yet this same book, and the logic of Aristotle, are the two most unobjectionable and unassailable treatises which ever were written.

The reason of the difficulty which is found in the Fifth Book is twofold. Firstly;—It is all reasoning, unhelped by the senses: most of the propositions have no portion of that intrinsic evidence which is seen in " two sides of a triangle are greater than the third;" but, at the same time, the propositions of arithmetic which correspond to those of the Fifth Book are very evident, and the student is therefore led to escape from the notion of *magnitude,* and fly to that of *number.* Secondly;—The non-existence of any very easy notation and system of arithmetic in the time of Euclid, made geometrical considerations relatively so much more simple, that the form of his book is (to us) unnecessarily remote from all likeness to a treatise connected with numbers. The difference between our day and his lies in this: that in the former the exactness of geometry was gained with some degree of prolixity and (to a beginner) obscurity; in the latter, the facility of arithmetic is preferred, and perfect demonstration is more or less sacrificed to it. I shall now endeavour to present the Fifth Book of Euclid in a form which will be more easy than the original, to those who have some acquaintance with algebra.

By *number* is here meant what is called *abstract* number, which merely conveys the notion of times or repetitions, considered independently of the things counted or repeated. By *magnitude,* or *quantity,* is meant a thing presented to us, not as to its form, if it have form, or as to colour, weight, or any other circumstance, but simply as that which is made up of parts, not differing from the whole in any thing but in being less; so that, if we consider separately a part and the whole, we have only two inferences:

The part is less than the whole.

The whole is greater than the part.

Every thing we can see or feel presents to us the notion of magnitude or quantity. And here we must observe, that we have to pick our words from among those in common use, which never have very precise meanings. For instance, we have *magnitude,* the nearest English word to which is *greatness;* and *quantity,* for which the word, if it existed, should be *so-much-ness.* These words are of the same meaning, and the more indefinite we now leave them (except only in assigning that they are to be considered as applied to any thing which can be made *more* or *less*), the better for our purpose; since it is the object of this treatise to deduce from that indefinite notion a method of making mathematical comparisons of quantities, by aid of the notion of number.

Upon two magnitudes, our senses will enable us to draw one or other of the following conclusions:

1. The first is sensibly *greater* than the second.
2. The first is sensibly *less* than the second.

3. The first is sensibly *equal* to the second; meaning that the difference, if any, is so small that our senses cannot perceive it. This is what is meant by equality of magnitudes in common life. The English foot and the Florence foot are equal for common purposes: they differ by about the twentieth part of an inch, which in a foot is called nothing.

Perfect equality is a mathematical conception, which never can be absolutely verified in practice; for so long as the senses cannot perceive a certain quantity, be it ever so small, so long it must always be possible that two quantities, which appear equal, may differ by as much as the imperceptible quantity. But we are not reasoning upon what we can carry into effect, but upon the conceptions of our own minds, which are the exact limits we are led to imagine by the rough processes of our hands. The following, then, is the postulate upon which we construct our results:

Any one magnitude being given, let it be granted that any number of others may be found, each of which is (positively and mathematically) equal to the first.

Let A represent a magnitude—not as in algebra, the number of units which it contains, but the magnitude itself—so that if it be, for instance, weight of which we are speaking, A is not a number of pounds, but the weight itself. Let B represent another magnitude of the same kind; we can then make a third magnitude, either by putting the two magnitudes together, or by taking away from the greater a magnitude equal to the less. Let these be represented by $A+B$ and $A-B$, A being supposed the greater. We can also construct other magnitudes, by taking a number of magnitudes each equal to A, and putting any number of them together. Thus we have

$A+A$	which abbreviate into	$2A$
$A+A+A$	....	$3A$
$A+A+A+A$	....	$4A$

and so on. We have thus a set of magnitudes, depending upon A, and all known when A is known; namely,

$$A \quad 2A \quad 3A \quad 4A \quad 5A \quad \&c.$$

which we can carry as far as we please. These (except the first) are distinguished from all other magnitudes by the name of *multiples* of A; and it is evident that they increase continually. Let the preceding be called the *scale of multiples* of A. It is clear that the multiples of multiples are multiples; thus, 7 times 3 A is 21 A, m times nA is (mn)A, where mn is the arithmetical product of the whole numbers m and n. The following propositions may then be proved.

Prop. I. If A be made up of B and C, then any multiple of A is made up of the same multiples of B and C; for 2 A must be made up of

B C B C

of which B and B make 2 B, C and C make 2 C; so that 2 A is made up of 2 B and 2 C. Similarly, 3 A is made up of

B C B C B C

or of 3 B and 3 C.

Corollary. Hence it follows, that if A be less than B by C, any multiple of A is less than the same multiple of B by the same multiple of C. For, since A is less than B by C, A and C together make up B; therefore, 2 A and 2 C make up 2 B, or 2 A is less than 2 B by 2 C. The algebraical representations of these theorems are as follows:

If $\quad A = B + C \qquad mA = mB + mC$

If $\quad A = B - C \qquad mA = mB - mC$

m being any of the numbers 2, 3, 4, &c.......

Prop. II. However small A may be, or however great B may be, the multiples in the scale

A, 2 A, 3 A, 4 A, 5 A, &c.

will come in time to exceed B, by continuing the scale sufficiently far: B and A being magnitudes of the same kind. This is a proposition which must be considered as self-evident: it must be remembered that B remains the same, while we pass from one multiple of A to the next. Put feet together and we shall come in time to exceed any number of miles, say a thousand. But the best illustration of the reason why we formally put forward so self-evident a proposition, will be to remark, that it is not *every* way of adding magnitude to magnitude without end, which will enable us to surpass any given magnitude. To a magnitude add its half; to that sum add half of the half; to which add the half of the last: and so on. No continuation of this process, were it performed a hundred million of times, could ever double the first magnitude.

Prop. III. If A be greater than B, any multiple of A is greater than the same multiple of B. This follows from Prop. I. And if A be less than B, any multiple of A is less than the same multiple of B. This follows from the corollary, Prop. I. And if A be equal to B, any multiple of A is equal to the same multiple of B. This is self-evident.

Prop. IV. If any multiple of A be greater than (equal to, or less than) the same multiple of B, then A is greater than (equal to, or less than) B. For example, let 4 A be greater than 4 B; then A must be greater than B; for, if not, 4 A would be equal to, or less than, 4 B (Prop. III.).

Prop. V. If from a magnitude the greater part be taken away; and if from the remainder the greater part of itself be taken away, and so on: the given magnitude may thus be made as small as we please, meaning as small as, or smaller than, any second magnitude we choose to name.

Let A and Z be the two magnitudes, and let A diminished by more than its half be B, then 2 B is less than A. Let B diminished by more than half be C; then 2 C is less than B, 4 C is less than 2 B, and still more less than A. Let C diminished by more than its half be D, then 2 D is less than C, 8 D is less than 4 C, and still more than A. This process must end by bringing one of the quantities A, B, C, D, &c. below Z in magnitude. For, if not, let A, B, C, &c. always remain greater than Z. Then, since 2 B, 4 C, 8 D, 16 E, &c. are all less than A (just proved) still more must 2 Z, 4 Z, 8 Z, 16 Z, &c. be less than A. But this cannot be; therefore, one of the set A, B, C, &c. must be less than Z.

[The *reductio ad absurdum*, as this sort of argument is usually called, is a difficult form of a simple inference. Suppose it proved that whenever P is Q, then X is Y. It follows that whenever X is *not* Y, P is not Q. It is usually held enough to say, for if P *were* Q X *would be* Y. But the form in which Euclid argues, supposes an opponent; and the whole argument then stands as follows. "When X is Y, you grant that P is Q; but you grant that P is not Q. I say that X is not Y. If you deny this you must affirm that X *is* Y, of which you admit it to be a consequence that P *is* Q. But you grant that P is not Q; therefore, you say at one time that P is Q and that P is not Q. Consequently, one or other of your assertions is wrong, either 'P is not Q' or 'X is Y.' If the first be right, the second is wrong: that is, 'X is not Y' is right."

The preceding argument runs as follows;—when A, B, C, &c. are all greater than Z, then 2 Z, 4 Z, &c. are all less than A: but 2 Z, 4 Z, &c. are not all less than A; therefore, A, B, C, &c. are not all greater than Z].

Corollary. The preceding proposition is equally true when, instead of taking more than the half at each step, we take the half itself in some or all of the steps.

Prop. VI. If there be two magnitudes of the same kind, A and B, and if the scales of multiples be formed

$$\text{A}, \quad 2\text{A}, \quad 3\,\text{A}, \quad \&\text{c.} \qquad \text{B}, \quad 2\,\text{B}, \quad 3\,\text{B}, \quad \&\text{c.}$$

then one of these two things must be true; EITHER, there are multiples in the first scale which are equal to multiples in the second scale; OR, there are multiples in the first scale which are as nearly equal as we please to multiples (not the same perhaps) in the second set: that is, we can find one of the first set, say mA, which shall either be equal to another in the second set, say nB, or shall exceed or fall short of it by a quantity less than a given quantity Z, which we may name as small as we please.

Let us take a multiple out of each set, any we please, say pA and qB. If pA and qB be equal, the first part of the alternative exists; if not, one must exceed the other. Let pA exceed qB, say by E; then we have

$$p\text{A} = q\text{B} + \text{E} \quad \ldots\ldots\ldots\ldots \quad (1)$$

Now E is either less than B, or equal to B, or greater than B. If the first, let it remain for the present; if the second, we have $p\text{A} = (q+1)\,\text{B}$, or the first alternative exists: if the third, then B can be so multiplied as to exceed E. Let $(t+1)\,\text{B}$ be the *first* multiple of B which exceeds E; that is, let the next below, or tB, be less than E, say by G, then we have

$$\text{E} = t\text{B} + \text{G} \qquad\qquad p\text{A} = q\text{B} + t\text{B} + \text{G}$$

or

$$p\text{A} = (q+t)\,\text{B} + \text{G}$$

Now G must be less than B; for E or $t\text{B} + \text{G}$ is less than $(t+1)\,\text{B}$, or $t\text{B} + \text{B}$. We have then made this first step (observe that $q + t$ is only *some* multiple of B; call it rB). Either the first alternative exists, or we can find pA and rB, so that

$$pA = rB + G \text{ where } G \text{ is less than } B \;\ldots\ldots\ldots\; (2)$$

Now G can be so multiplied as to exceed B; let vG and $(v+1)G$ be the multiples of G, between which B lies, so that

$$\begin{aligned} vG &\text{ is less than } B, \text{ say } & vG &= B - K \\ (v+1)G &\text{ is greater than } B, \text{ say } & vG + G &= B + L \end{aligned}$$

and it follows that $K + L = G$, for since $(v+1)G$ and vG differ by G, if a magnitude lie between them, their difference must be made up of the excess of that magnitude over the lesser, together with its defect from the greater. Consequently, either K and L are both halves of G, or one of them falls short of the half. Suppose K is less than the half of G: then take both sides of (2), v times, and we have

$$vpA = vrB + vG$$

or

$$vpA = vrB + B - K$$

or

$$vpA = (vr+1)B - K \text{ (K less than half G)}$$

But if L be less than the half of G, take both sides of (2) $v+1$ times which gives

$$(v+1)pA = (v+1)rB + (v+1)G$$

or

$$(v+1)pA = \overline{v+1}\cdot rB + B + L$$

or

$$\overline{v+1}\; pA = (\overline{v+1}\,r + 1)B + L \text{ (L less than half G)}$$

If K and L are both halves of G, we may take either. And if (a case not yet included) a multiple of G, vG, be exactly equal to B, we have then

$$vpA = vrB + vG = (vr+1)B$$

which gives the first alternative. Consequently, we either prove the first alternative, or we reduce the equation

$$pA = rB + G \quad \text{(G less than B)}$$

to an equation of the form

$$p'A = r'B \pm G' \;\{G' \text{ not greater than the half of } G.$$

We may now proceed as before; but, to exemplify all the cases that may arise, let us take

$$p'A = r'B - G'$$

If $v'G'$ be exactly B, we prove the first alternative, as before; but if B lie between $v'G'$ and $(v'+1)G'$, let us suppose

$$\left.\begin{aligned} v'G' &= B - K' \\ (v'+1)G' &= B + L' \end{aligned}\right\} \text{ and } K' + L' = G' \text{ as before,}$$

in which one of the two, K' or L', will not be greater than the half of G', so that we obtain by the same process, an equation of the form

$$p''A = q''A \pm G'' \left.\right\} \begin{array}{l} G'' \text{ not greater than} \\ \text{the half of } G'. \end{array}$$

By proceeding in this way, we prove either, 1. The first alternative of the proposition; or, 2. the possibility of forming a continued set of equations

$$pA = qB \pm G, \quad p'A = q'B \pm G', \quad p''A = q''B \pm G'', \text{ \&c.}$$

where, in the scale of quantities G, G', G'', &c., no one exceeds the half of the preceding. Consequently, we may (unless interrupted by the first alternative) carry on this process until one of the quantities G, G', G'' &c. is smaller than Z (Prop. V.) that is, we have either the first or second alternative of the problem. And exactly the same demonstration may be applied to the case, where at the outset $pA = qB - E$.

This proposition proves nothing of a single magnitude, but it establishes two apparently very distinct relations between magnitudes considered in pairs. There may be cases in which the first alternative is established at last: and there may be cases in which it is never established. We shall first take the case in which the first alternative is established.

Suppose it ascertained by the preceding process that

$$8A = 5B$$

Here is an arithmetical equation between the magnitudes: and therefore any processes of *concrete* arithmetic will apply. Take the 40th part ($8 \times 5 = 40$) of both sides,

which gives $\dfrac{8A}{40} = \dfrac{5B}{40}$ or $\dfrac{A}{5} = \dfrac{B}{8}$

consequently the fifth part of A is the same as the eighth part of B, or that which is contained 5 times in A is also that which is contained

8 times in B. Let this fifth of A or eighth of B be called M; then $A=5M$, $B=8M$, and A and B are both multiples of M. Consequently, when the first alternative of Prop. VI. exists, both A and B are multiples of some third magnitude M. The converse is readily proved, namely, that when A and B are both multiples of any third magnitude, the first alternative of Prop. VI. is true. For if $A=xM$, $B=yM$, we have $yA=yxM$, $xB=xyM$, or $yA=xB$. The term *measure* is used conversely to multiple, thus: if A be a multiple of M, M is said to be a measure of B. Hence in the case we are now considering, A and B have a *common measure*, and are said to be *commensurable*. We have therefore shewn that all commensurable magnitudes, and commensurable magnitudes only, satisfy this first alternative.

There remains, then, only the second case to consider, which it is now evident contains those magnitudes (if any such there be) which have no common measure whatsoever. The question therefore is, Are there such things as incommensurable magnitudes? On this point the second alternative shews that our senses cannot judge, for let Z be the least magnitude of the kind in question, which they are capable of perceiving (of course with the best telescopes, or other means of magnifying small quantities which can be obtained) then we know that pA may be made to differ from qB by less than Z, that is, we may say that all magnitudes are *sensibly* commensurable. But it evidently does not follow that all magnitudes are mathematically commensurable; and it has been shewn, by process of demonstration, that there are* incommensurable quantities in such abundance, that take almost any process of geometry we please, the odds are immense against any two results being commensurable.

The suspicion that all magnitudes must be commensurable led to the attempt, which lasted for centuries, to find the exact ratio of the circumference of a circle to its diameter. And even now, though the adventure is never tried by those who have knowledge enough to read demonstration of its impossibility, no small number of persons

* Legendre, and others before him, have shewn that the diameter and circumference of a circle are incommensurable; and the student will find in my Algebra, p. 98, or in the *Lib. Useful Know.*, treatise on the *Study of Mathematics*, p. 81, proof that the side and diagonal of a square are incommensurables. Also in Legendre's Geometry, or Sir D. Brewster's Translation.

exercise themselves by endeavouring to make an elementary acquaintance with geometry (and sometimes none at all) overcome this difficulty. It is our business here to shew how strict deductions may be made upon quantities which are incommensurable, with the same facility, and in the same manner, as upon commensurables. If we call any length (say that known by the name of a foot), the unit of its kind, and denote it in calculation by 1, we must call twice such a magnitude 2, and so on; half such a magnitude ½, and so on. We may then apply arithmetic, every possible subject of which is contained in the following infinitely extended table.

0				1				2	&c.
0		$\frac{1}{2}$		1		$\frac{3}{2}$		2	&c.
0	$\frac{1}{3}$	$\frac{2}{3}$		1	$\frac{4}{3}$	$\frac{5}{3}$		2	&c.
0	$\frac{1}{4}$	$\frac{2}{4}$	$\frac{3}{4}$	1	$\frac{5}{4}$	$\frac{6}{4}$	$\frac{7}{4}$	2	&c.
	&c.		&c.		&c.		&c.		

And every length which is commensurable with the foot is included, in many different forms, in this table. Let F represent the foot, L any other length, let M be their common measure: let

$$F=fM,\quad L=lM,\ \text{then}\ lF=fM,\ \text{or}$$

$$L=\frac{l}{f}F=\frac{l}{f}\ \text{when F is called 1.}$$

But it is plain that we cannot, by any arithmetic of length founded upon the foot as a unit, draw conclusions as to lengths which are incommensurable with the foot, though we can perhaps do so for any practical purpose. Let L be a length which is such, and let Z be a length so small as to be immaterial for the purpose in question. Then, we can determine l and f, so that

$$fL=lF\pm G \qquad \text{(G less than Z)}$$

or

$$L=\frac{l}{f}F\pm\frac{G}{f}:$$

so that, by assuming $L=\frac{l}{f}F$, we commit an error, in excess or defect, less than G, and therefore immaterial. With such a process many minds would rest contented; but there is a consideration which will stand in the way of perfect satisfaction, or at least ought to do so. Granting that in the preceding case the error at the outset is immaterial, let us suppose the student disposed to substitute for all incommensurables, magnitudes very near to them which are commensurables, and thus to continue his career till he comes to the highest branches of applied mathematics. Let us suppose a set of processes, beginning in arithmetic, continued through algebra, the differential calculus, &c., up to a point in optics or astronomy, in a series of results, embracing, we may suppose, ten thousand inferences. If he set out with an erroneous method, what security has he that the error will not be multiplied ten thousand fold at the end, and thus become of perceptible magnitude. If somebody acquainted with the subject have told him that it will not so happen, he might as well skip the intermediate sciences and receive the result he wants to obtain on the authority of that person, as study them in a manner, the correctness or incorrectness of which depends on that person's authority. If he answer that the result, namely, such multiplication of errors, appears extremely improbable, it may be replied, *firstly*, that that is more than he can undertake to decide; *secondly*, that by pursuing his mathematical studies on such a presumption, he makes all the pure sciences present *probable* results only, not demonstrated results; more probable, perhaps, than many parts of history, but resting on an impression which must, in his mind, be the result of testimony.

It appears, however, that we may expect series of collateral results, the one for commensurables, the other for incommensurables, and presenting great resemblances to each other; for we may, by any alteration, however minute, convert the latter kind of magnitude into the former. But this we may prevent, by extending our notions of arithmetical operations, or rather by applying to *magnitude* processes which are usually applied to *number* only, as follows:

If we examine the processes of arithmetic, we find, 1st, Addition and substraction, to which abstract number is not necessary, since the concrete magnitudes themselves can be added or subtracted. 2d, Multiplication, the raising of powers and the extraction of roots, in all of which abstract number is essentially supposed to be the subject of operation. 3d, Division, in which it is not necessary to suppose abstract number in finding the whole part of the quotient, but in which we cannot, without reference to numbers, compare the remainder and divisor, in order to form the finishing fraction of the quotient.

4th, The process of finding the greatest common measure of two quantities, in which the remainder is not compared with the divisor, except in a manner which is as applicable to the case of concrete magnitudes as of abstract numbers. To shew this, we shall demonstrate the method of finding the greatest common measure of two *magnitudes*.

Let A and B be two magnitudes, which have a common measure M; let $A = aM$, $B = bM$. Then, it is clear that

$$xA + yB \text{ or } (xa + yb)M, \quad xA - yB \text{ or } (xa - yb)M$$

have the same measure, unless it should happen that in the latter case $xa = yb$, in which case $xA = yB$. Let A be the greater of the two, and let A contain B more than β and less than $\beta + 1$ times, so that $A = \beta B + B'$, when B' is less than B. Then B' being $A - \beta B$, is measured by M. Let B contain B' more than β' and less than $\beta' + 1$ times; or let $B = \beta' B' + B''$ where B'' is less than B'. Let B' contain B'' more than β'' times, &c., or let $B' = \beta'' B'' + B'''$, and so on. And B'' or $B - \beta' B'$ is measured by M, &c. We have then the following conditions:

	A is a multiple of	M
	B	M
$A = \beta B + B'$	$B' < B$,	but is a multiple of M
$B = \beta' B' + B''$	$B'' < B'$	....
$B' = \beta'' B'' + B'''$	$B''' < B''$	....

Now, since B B′ B″ are decreasing quantities, and all multiples of M, they are all to be found in the series,

$$M, \qquad 2M, \qquad 3M, \qquad 4M, \qquad \&c.$$

in which continual decrease must bring us at last to nothing, or we must end with an equation of the form.*

$$B^{(n)} = \beta^{(n+1)} B^{(n+1)} + 0$$

that is, one remainder is a multiple of the next. To take a case, let the fifth equation finish the process; so that, in addition to the preceding, we have

$$B'' = \beta''' B''' + B^{iv}$$
$$B''' = \beta^{iv} B^{iv}$$

* When a letter denotes an indefinite number of accents, it is distinguished from an exponent by being placed in brackets, and higher numbers of accents than three are usually denoted by *Roman* numerals.

In the fourth, substitute B‴ from the fifth, giving

$$B'' = (\beta''' \beta^{iv} + 1) B^{iv}$$

In the third, substitute B″ and B‴ as found, giving

$$B' = (\beta'' \beta''' \beta^{iv} + \beta'' + \beta^{iv}) B^{iv}$$

In the second, substitute B′ and B″ as found, giving

$$B = (\beta' \beta'' \beta''' \beta^{iv} + \beta' \beta'' + \beta' \beta^{iv} + \beta''' \beta^{iv} + 1) B^{iv}$$

In the first, substitute B and B′ as found, giving

$$A = (\beta \beta' \beta'' \beta''' \beta^{iv} + \beta \beta' \beta'' + \beta \beta' \beta^{iv} + \beta \beta''' \beta^{iv} + \beta'' \beta''' \beta^{iv} + \beta + \beta'' + \beta^{iv}) B^{iv}$$

Consequently, B^{iv} is a common measure of A and B; but, since M at the outset is any common measure we please, let it be the greatest common measure. Then B^{iv} must be M, for it is in the series M, 2 M, &c.; and were it any other than M, there would be B^{iv} a common measure, greater than the greatest. Hence this process determines the greatest common measure, and also the number of times which each of the two, A and B, contains the greatest common measure.

It is here most essential to observe, that this whole process is independent of any arithmetic, except pure addition and subtraction, which can be performed on the magnitudes themselves, without any numerical relation whatsoever; the only thing required being the axiom in page 3. We shall actually exemplify this on two right lines.

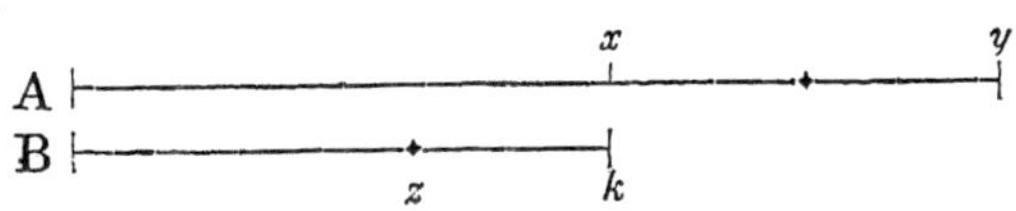

$$A = B + (xy) \quad B = (xy) + (zk) \quad (xy) = 2(zk)$$

Therefore $B = 3(zk) \qquad A = 5(zk)$

In this case, by actual measurement (supposed geometrically exact) B and A are found to be respectively 3 and 5 times zk.

When the preceding process has an end, we therefore detect the greatest common measure; and we have shewn, that where there is a common measure, the process must give it, with the converse. Consequently, in the case where there is no common measure, this process must go on for ever, and we have an interminable series of equations, $A = \beta B + B'$, $B = \beta' B' + B''$, $B' = \beta'' B'' + B'''$, &c., the conditions of which are, that B, B′, B″, B‴, &c., are a continually decreasing series, though it does not follow that each one was less than the half of the preceding. We shall now examine the effect of successive substitutions from the beginning, first making the following remark: If there be any two incommensurable quantities A and B, of which A is the greater, then there follows an interminable set of whole numbers, $\beta, \beta', \beta'', \ldots.$ which are not subject to any particular law, but can be found when A and B are given; and an interminable set of quantities, A, B, B′, B″, ... connected with the former by this law, that A contains B between β and $\beta+1$ times; B contains B′ between β' and $\beta'+1$ times, and so on.

We have $B' = A - \beta B$

$$B'' = B - \beta' B' = B - \beta'(A - \beta B)$$

or $B'' = (\beta\beta' + 1)B - \beta' A$

$$B''' = B' - \beta'' B'' = A - \beta B - (\beta\beta' + 1)\beta'' B + \beta'\beta'' A$$
$$= (\beta'\beta'' + 1)A - (\beta\beta'\beta'' + \beta + \beta'')B$$

and thus we go on representing the remainders alternately, in the form $pA - qB$ and $qB - pA$. We may easily find the law of the coefficients, as follows:

Suppose we come to

$$B^{(n)} = qB - pA$$
$$B^{(n+1)} = p'A - q'B$$

Then we have $B^{(n)} = \beta^{(n+1)} B^{(n+1)} + B^{(n+2)}$

or $B^{(n+2)} = B^{(n)} - \beta^{(n+1)} B^{(n+1)}$

$$= qB - pA - \beta^{(n+1)}(p'A - q'B)$$
$$= (\beta^{(n+1)} q' + q)B - (\beta^{(n+1)} p' + p)A$$

or if, continuing the preceding notation, we suppose

$$B^{(n+2)} = q''B - p''A$$

we have $p'' = \beta^{(n+1)} p' + p \qquad q'' = \beta^{(n+1)} q' + q$

so that, if we write the values of B′, B″, ... with the following notation, putting opposite to each the $\beta^{(n)}$ which occurs *for the first time* in the $B^{(n)}$ of the equation; namely,

$$B' = p_1 A - q_1 B \quad \ldots\ldots\ldots\ldots\ldots \quad \beta$$
$$B'' = q_2 B - p_2 A \quad \ldots\ldots\ldots\ldots\ldots \quad \beta'$$

$$
\begin{array}{lll}
B''' = p_3 A - q_3 B & \ldots\ldots\ldots\ldots\ldots & \beta'' \\
B^{iv} = q_4 B - p_4 A & \ldots\ldots\ldots\ldots\ldots & \beta''' \\
\&c. \quad \&c. & & \&c.
\end{array}
$$

we have the following uniform method of forming p_n and q_n for different values of n in succession.

$$
\begin{array}{ll}
p_1 = 1 & q_1 = \beta \\
p_2 = \beta' & q_2 = \beta'\beta + 1 \\
p_3 = \beta'' p_2 + p_1 & q_3 = \beta'' q_2 + q_1 \\
p_4 = \beta''' p_3 + p_2 & q_4 = \beta''' q_3 + q_2 \\
\&c. \quad \&c. & \&c. \quad \&c.
\end{array}
$$

in which it is plain, from the method of formation, that p_1 p_2 &c. q_1 q_2 &c. are increasing whole numbers, so that we may continue, supposing B′ B″ .... never fail, till p_n and q_n are greater than any number named. And since B′ B″ .... are all less than B, and therefore less than A, we have the following succession of results, *ad infinitum*.

$$
\begin{array}{llll}
p_1 A \text{ is greater than } & q_1 B & \text{but less than } & (q_1 + 1) B \\
p_2 A \text{ is less than } & q_2 B & \text{but greater than } & (q_2 - 1) B \\
p_3 A \text{ is greater than } & q_3 B & \text{but less than } & (q_3 + 1) B \\
\&c. & \&c. & & \&c.
\end{array}
$$

Hence, it appears that

$$
\begin{array}{rl}
A \text{ is greater than } & \frac{q_1}{p_1} B \\
\text{less than } & \frac{q_2}{p_2} B \\
\text{greater than } & \frac{q_3}{p_3} B \\
\text{less than } & \frac{q_4}{p_4} B \text{ \&c. } \textit{ad inf.}
\end{array}
$$

Now, from this table of relations, we can determine whether any given multiple of A, xA, is greater or less than any given multiple of B, yB. To do this we must inquire between what two consecutive multiples of B does xA lie.

We now proceed as follows:

1. We must shew that any fraction, such as

$$\frac{a+m}{b+n} \text{ lies between } \frac{a}{b} \text{ and } \frac{m}{n}$$

unless where the two latter are equal, in which case the first also is the same. The preceding must be true if

$$a + m \text{ lies between } \frac{a}{b}(b+n) \text{ and } \frac{m}{n}(b+n)$$

$$\text{or} \qquad a + \frac{an}{b} \text{ and } \frac{bm}{n} + m$$

$$\text{or} \qquad a + m + \frac{an}{b} - m \text{ and } a + m + \frac{bm}{n} - a$$

$$\text{or} \qquad a + m + n\frac{a}{b} - n\frac{m}{n} \text{ and } a + m + b\frac{m}{n} - b\frac{a}{b}$$

which is evidently true: for if $\frac{a}{b}$ be greater than $\frac{m}{n}$, the first is greater than $a+m$, and the second less; if $\frac{a}{b}$ be less than $\frac{m}{n}$, *vice versâ.*

2. We now see that

$$\frac{q_3}{p_3} \text{ or } \frac{\beta'' q_2 + q_1}{\beta'' p_2 + p_1} \text{ lies between } \frac{\beta'' q_2}{\beta'' p_2} \text{ and } \frac{q_1}{p_1}$$

$$\text{or} \quad \frac{q_2}{p_2} \text{ and } \frac{q_1}{p_1}$$

$$\frac{q_4}{p_4} \text{ or } \frac{\beta''' q_3 + q_2}{\beta''' p_3 + p_2} \ldots\ldots\ldots\ldots \frac{q_3}{p_3} \quad \frac{q_2}{p_2}$$

and so on. Consequently, to arrange all the fractions thus considered, in order of magnitude, we must write them thus,

$$\frac{q_1}{p_1} \quad \frac{q_3}{p_3} \quad \frac{q_5}{p_5} \ldots \qquad \ldots \frac{q_6}{p_6} \quad \frac{q_4}{p_4} \quad \frac{q_2}{p_2}$$

3. We can thus bring two fractions as near together as we please: to prove this, take three consecutive fractions

$$\frac{q_m}{p_m} \quad \frac{q_{m+1}}{p_{m+1}} \quad \left(\frac{q_{m+2}}{p_{m+2}} \text{ or } \frac{\beta^{(m+1)} q_{m+1} + q_m}{\beta^{(m+1)} p_{m+1} + p_m}\right)$$

which reduced to a common denominator, the first and second, and the second and third, give

$$\frac{q_m\, p_{m+1}}{p_m\, p_{m+1}} \qquad \frac{q_{m+1}\, p_m}{p_{m+1}\, p_m}$$

$$\text{and} \quad \frac{\beta^{(m+1)} q_{m+1}\, p_{m+1} + p_m\, q_{m+1}}{p_{m+1}\, p_{m+2}} \quad \text{and} \quad \frac{\beta^{(m+1)} q_{m+1}\, p_{m+1} + p_{m+1}\, q_m}{p_{m+1}\, p_{m+2}}$$

in which it is clear that the difference of the numerators is the same in each couple, but that if the first numerator be the greater of the first couple, the second numerator is that of the second; a result we might have foreseen, having proved that

$$\frac{q_{m+2}}{p_{m+2}} \text{ lies between } \frac{q_{m+1}}{p_{m+1}} \text{ and } \frac{q_m}{p_m}.$$

Hence it follows that the numerator of the difference of any two successive fractions of the set

$$\frac{q_1}{p_1} \quad \frac{q_2}{p_2} \quad \frac{q_3}{p_3} \quad \ldots\ldots\ldots\ldots$$

is the same as that of the difference immediately preceding, that is,

the difference of $\frac{q_n}{p_n}$ and $\frac{q_{n-1}}{p_{n-1}}$ has the same numerator as the difference of $\frac{q_{n-1}}{p_{n-1}}$ and $\frac{q_{n-2}}{p_{n-2}}$ ...... which has the same numerator as the difference of $\frac{q_2}{p_2}$ and $\frac{q_1}{p_1}$; but

$$\frac{q_1}{p_1} - \frac{q_2}{p_2} = \frac{1}{\beta} - \frac{\beta'}{\beta'\beta+1} = \frac{1}{\beta(\beta\beta'+1)}$$

therefore this numerator of the differences is always 1, or

$$\frac{q_n}{p_n} \text{ and } \frac{q_{n+1}}{p_{n+1}} \text{ differ by } \frac{1}{p_n\, p_{n+1}}$$

Hence the difference may be made as small as we please, or smaller than any fraction $\frac{1}{m}$ named by us, since p_n itself can be made greater than m, much more $p_n\, p_{n+1}$.

4. These fractions cannot for ever lie alternately on one side and the other of any *given* fraction $\frac{v}{x}$.

For if this were possible, then, since A lies between

$$\frac{q_n}{p_n}\text{B} \quad \text{and} \quad \frac{q_{n+1}}{p_{n+1}}\text{B}$$

and since by the supposition $\frac{v}{x}$B does the same, and since the couple just mentioned can be made to differ by as small a fraction of B as we please, then we should have

$$\text{A} = \frac{v}{x}\text{B} \pm \text{K}$$

where K may be made as small as we please. Now this is saying that $A = \frac{v}{x}B$; for A must either be

$$\frac{v}{x}B \quad \text{or} \quad \frac{v}{x}B \pm \text{ some definite magnitude;}$$

but the latter it is not; for the supposition we are trying leads to

$$A = \frac{v}{x}B + \text{ a magnitude as small as we please.}$$

Consequently, our supposition that the series of fractions lie alternately on one side and the other of a definite fraction $\frac{v}{x}$, leads to the conclusion that A and B are commensurable, or the process of finding B′ B″ .... finishes, as we have shewn. But it does not finish, by hypothesis; therefore the series of fractions cannot lie alternately on one side and the other of $\frac{v}{x}$.

We can now shew between what multiples of B xA must lie. It is clear that

$$xA \text{ lies between } \frac{x\,q_n}{p_n}B \text{ and } \frac{x\,q_{n+1}}{p_{n+1}}B:$$

now it is not possible that any whole number v should always lie between $\frac{x\,q_n}{p_n}$ and $\frac{x\,q_{n+1}}{p_{n+1}}$; for if so, then would

$$\frac{v}{x} \text{ always lie between } \frac{q_n}{p_n} \text{ and } \frac{q_{n+1}}{p_{n+1}}$$

which has been proved to be impossible. Consequently,

$$\frac{x\,q_n}{p_n}B \text{ and } \frac{x\,q_{n+1}}{p_{n+1}}B \quad \text{(which approach each other without limit)}$$

must come at last always to lie between two multiples of B; and still more must xA, which lies between *them*. Hence, by proceeding far enough, we can always find between what multiples of B lies xA; and thence whether xA is greater or less than yB.

We have thus divided all pairs of magnitudes into two classes,

1. *Commensurables*, in which we can always say that $A = \frac{q}{p}B$, q and p being whole numbers, and can always tell exactly by what fraction of A or B, xA exceeds or falls short of yB. For we have

$$xA - yB = \left(x - y\frac{p}{q}\right)A = \left(x\frac{q}{p} - y\right)B \quad \text{if} \quad xA > yB$$

$$(yB - xA) = \left(\frac{p}{q}y - x\right)A = \left(y - x\frac{q}{p}\right)B \quad \text{if} \quad xA < yB$$

2. *Incommensurables*, in which we can never say $A = \frac{q}{p}B$, but in which we can assign a series of fractions alternately increasing and decreasing, but making less and less change at every step,

$$\frac{q_1}{p_1} \quad \frac{q_2}{p_2} \quad \frac{q_3}{p_3} \quad \ldots\ldots$$

and such that A is greater than $\frac{q_1}{p_1}B$, less than $\frac{q_2}{p_2}B$, &c. *ad infinitum*: so that we can always assign

$$A = \frac{q_n}{p_n}B + K$$

where K is less than any magnitude we name; and such that we can always tell by them whether xA exceeds or falls short of yB, but not exactly *how much*.

Let us suppose, as an example, that we have two magnitudes A and B, which tried by the process in page 13, give

$A = B + B'$, $B = B' + B''$, $B' = B'' + B'''$, &c. *ad inf.*

or suppose $\beta = 1$, $\beta' = 1$, $\beta'' = 1$, &c. *ad inf.*

Hence the several values of p and q are $p_1 = 1$, $p_2 = 1$, $p_3 = 2$, &c. as in this table,

	1	2	3	4	5	6	7	8	9	10	11	12	&c.
p	1	1	2	3	5	8	13	21	34	55	89	144	
q	1	2	3	5	8	13	21	34	55	89	144	233	&c.

or $$A > B < 2B > \frac{3}{2}B < \frac{5}{3}B > \frac{8}{5}B < \frac{13}{8}B \quad \&c.$$

Hence A lies between B and 2B
2A .... 3B .. 4B
3A .... 4B .. 5B
4A .... 6B .. 7B
5A .... 8B .. 9B &c.

If we wish to know between what multiples of B 100A lies, we find

$$A > \frac{144}{89}B < \frac{233}{144}B;\ 100A > 161\frac{71}{89}B < 161\frac{116}{144}B$$

or 100 A lies between 161 B and 162 B.

We can thus form what we may call a *relative* multiple scale made by writing down the multiples of A, and inserting the multiples of B in their proper places; or *vice versâ*. In the instance just given the commencement of this scale is

B, A, 2 B, 3 B, 2 A, 4 B, 3 A, 5 B, 6 B, 4 A, 7 B, 8 B, 5 A, 9 B, &c.

which we may continue as far as we please by simple arithmetic. If the magnitudes in question be lines, we may represent this multiple scale as follows:

Measuring from O, the crosses mark off multiples of A, and the bars multiples of B. Thus

$$O1_1 = B \qquad O1_2 = 2B \qquad O1_3 = 3B \text{ \&c.}$$
$$O\times_1 = A \qquad O\times_2 = 2A \qquad O\times_3 = 3A \text{ \&c.}$$

We shall now proceed to some considerations connected with a multiple scale, for the purpose of accustoming the mind of the student to its consideration. We may imagine a scale like the preceding to be equivalent to an infinite number of assertions or negations, each one connected with the interval of magnitude lying between two multiples of B. Thus, the preceding scale contains the following list *ad infinitum*.

1.	Between	0	and	B	lies no multiple of A
2.	Between	B	and	2 B	lies A
3.	Between	2 B	and	3 B	lies no multiple of A
4.	Between	3 B	and	4 B	lies 2 A
&c.		&c.		&c.	&c.

Now, on this we remark, 1st, That the negatives of the above series, though they appear at first to prove nothing, yet in reality have each an infinite number of negative consequences. From the third assertion of the preceding list, namely, neither A, nor 2 A, nor 3 A, &c. lies between 2 B and 3 B, we immediately deduce all the following: A does not lie between 2 B and 3 B, nor between $\frac{2}{2}$ B and

$\frac{3}{2}$B, nor between $\frac{2}{3}$B and $\frac{3}{3}$B, nor between $\frac{2}{4}$B and $\frac{3}{4}$B, &c. &c.
2d, Observe that every affirmative assertion in the above includes a certain number of the affirmative ones which *precede*, and an infinite number of parts of the negative ones preceding and following. For instance, we find that 100 A lies between 161 B and 162 B, or A lies between $\frac{161}{100}$B and $\frac{162}{100}$B, that is between B and 2 B. Again, 2 A lies between $\frac{322}{100}$B and $\frac{324}{100}$B, or between 3 B and 4 B. Similarly, 3 A lies between $\frac{483}{100}$B and $\frac{486}{100}$B, or between 4 B and 5 B; and it might thus seem at first as if every affirmation made all the affirmations preceding its necessary consequences. But if we try 90 A by the preceding, we shall find that it lies between

$$\frac{14490}{100}\text{B and } \frac{14580}{100}\text{B or between } 144\text{B and } 146\text{B}$$

so that we can only affirm 90 A to lie either between 144 B and 145 B, or between 145 B and 146 B, but we do not (from this) know which. But we can say that 90 A does not lie between 146 B and 147 B, or between 147 B and 148 B, &c. The points at which any affirmation does not determine those preceding may be thus found. Let kA lie between lB and $(l+1)$B; or

$$\text{A lies between } \frac{l}{k}\text{B and } \frac{l+1}{k}\text{B}$$

$$m\text{A} \;\ldots\ldots\ldots\ldots\; \frac{ml}{k}\text{B and } \frac{m(l+1)}{k}\text{B}$$

If $\frac{ml}{k}$ and $\frac{m(l+1)}{k}$ lie between t and $t+1$, then mA lies between tB and $(t+1)$B: but if, in going from the first to the second, we pass through a whole number, or if ml, divided by k, gives a quotient t and remainder r, and $m(l+1)$, divided by k, gives a quotient $t+1$ and remainder r', then we have

$$\frac{ml}{k} = t + \frac{r}{k} \qquad \frac{m(l+1)}{k} = t + 1 + \frac{r'}{k}$$

or $\quad m = k - r + r' \quad$ or $\quad r + m = k + r'$

and in all cases where $r+m$ is greater than k, this condition can be fulfilled. The process may be shortened, by using instead of l the remainder arising from dividing l by k. Suppose, for instance, it is required to determine what preceding affirmatives are ascertained by the proposition 10A lies between 33B and 34B. We have then $l=33$, $k=10$, remainder of $l \div k = 3$.

$m=2$	$t=6$	$r=6$	2A lies between 6B and 7B
$m=3$	$t=9$	$r=9$	————
$m=4$	$t=13$	$r=2$	4A .... 13B and 14B
$m=5$	$t=16$	$r=5$	5A .... 16B and 17B
$m=6$	$t=19$	$r=8$	————
$m=7$	$t=23$	$r=1$	7A .... 23B and 24B
$m=8$	$t=26$	$r=4$	————
$m=9$	$t=29$	$r=7$	————

By proceeding thus, it will appear that there is no perceptible law regulating the places of A, 2A, ... among B, 2B, &c., derivable from the sole condition of kA lying between lB and $(l+1)$B. Nevertheless, it is easy to prove, that if all the rest of the relative scale be given from and after any given point, that the whole of the preceding part can then be determined. For, suppose kB to be the commencement of the part of the scale given, and let the place of mA be asked for, which precedes hA, the first multiple of A appearing in the scale. Multiply m by g, so that mg shall be greater than h. Then mgA appears in the portion of the scale given, say between wB and $(w+1)$B. Therefore

$$m\text{A lies between } \frac{w}{g}\text{B and } \frac{w+1}{g}\text{B}$$

and if $\frac{w}{g}$ and $\frac{w+1}{g}$ lie between t and $t+1$, the question is settled; but this must always be the case, if we include the case where $\frac{w}{g}$ or $\frac{w+1}{g}$ is itself a whole number.

From all that precedes, we draw the following conclusions:

1. Having given A and B, two incommensurable magnitudes of the same species (both lengths, both weights, &c.), we can assign, by a processe s embling that of finding the greatest common measure in arithmetic, the relative scale of multiples of A and B, which points out between what two multiples of B any given multiple of A lies, or *vice versâ*.

2. Any part of the beginning of this scale being deficient, we can construct it by means of the rest.

3. We can find a magnitude which shall be commensurable with A, differing from B by less than any magnitude we name; and can assign the fraction which it is of A.

Given the two magnitudes, their relative multiple scale is given; but when the scale is given, the two magnitudes are not given. For it is easily proved that there is an infinite number of couples of magnitudes which have the same scale with any given one. Let the scale of A and B be given; then will the scale of $\frac{p}{q}$A and $\frac{p}{q}$B be the same, where p and q are any whole numbers whatsoever.

For if $\quad kA$ lie between lB and $(l+1)B$

then $\quad k\frac{p}{q}A$ lies between $l\frac{p}{q}B$ and $(l+1)\frac{p}{q}B$

or making $\quad \frac{p}{q}A = A' \qquad \frac{p}{q}B = B'$

kA' lies between lB' and $(l+1)B'$

whence the scale of A and B is the same as that of A′ and B′ for any value of k.

What is it, then, which is *given when the scale is given?* Not the magnitudes themselves; for if the scale belong to A and B, it also belongs to every one of the infinite cases of $\frac{p}{q}$A and $\frac{p}{q}$B. The scale, therefore, only defines such a relation between the magnitudes as belongs to 2 A and 2 B, 3 A and 3 B, &c., as well as to A and B. It is usual to call this relation the *proportion* between the two quantities in common life, and in mathematics their *ratio;* in Euclid the term is λόγος.

Two magnitudes, A and B, are said to have the same ratio as two other magnitudes, P and Q, when the relative scales of the two are the same; that is, when the multiples of Q are distributed as to magnitude among those of P, in the same way precisely as those of B are distributed among those of A. And P and Q may be two magnitudes of one kind, two areas, for instance, while A and B may be of another, two lines, for instance.

It is easy to shew that this accordance of scales is equivalent to the common idea of proportion, such as it would become if we took all means of comparison away, except that of multiples. Let us imagine A and B to be two lines in a picture, and P and Q the two corresponding lines in what is meant for an exact copy on a larger scale. Set an artist to determine whether P and Q are in the proper proportion to each other, without any assistance except the means of repeating A, B, P, Q, as many times as he pleases. He will reason as follows: "If Q be ever so little out of proportion to P, though it may not be visible to the eye, yet every multiplication of the two will increase the error, so that at last it will become perceptible. If there be a line 100A laid down in the first picture, and if it be found to lie between 51B and 52B, then should 100P lie between 51Q and 52Q. But if Q be a little wrong, then 100P may not lie between 51Q and 52Q."

It only remains to see whether this definition of proportion will include the case of commensurable quantities. These satisfy such an equation as $k\mathrm{A} = l\mathrm{B}$, k and l being two whole numbers, and it is easy to shew that the whole relative scale is divided into an infinite succession of similar portions. Firstly, this one equation determines the whole scale; for we have

$$\mathrm{A} = \frac{l}{k}\mathrm{B} \qquad\qquad m\mathrm{A} = \frac{ml}{k}\mathrm{B}$$

or if $\frac{ml}{k}$ lie between t and $t+1$, $m\mathrm{A}$ lies between $t\mathrm{B}$ and $(t+1)\mathrm{B}$

$$\text{if} \quad \frac{ml}{k} = t \qquad\qquad m\mathrm{A} = t\mathrm{B}$$

Let us suppose, for instance, $\mathrm{A} = \frac{7}{4}\mathrm{B}$. Then we have

A	lies between	B	and	2B
2A	....	3B	and	4B
3A	....	5B	and	6B
4A	is equal to	7B :	or the scale is	

$$0 \quad \mathrm{B} \quad \mathrm{A} \quad 2\mathrm{B} \quad 3\mathrm{B} \quad 2\mathrm{A} \quad 4\mathrm{B} \quad 5\mathrm{B} \quad 3\mathrm{A} \quad 6\mathrm{B} \quad \begin{matrix} 7\mathrm{B} \\ 4\mathrm{A} \end{matrix}$$

From this point the scale begins again in the same order. Thus, the second portion is

$$\begin{matrix} 7\mathrm{B} \\ 4\mathrm{A} \end{matrix} \quad 8\mathrm{B} \quad 5\mathrm{A} \quad 9\mathrm{B} \quad 10\mathrm{B} \quad 6\mathrm{A} \quad 11\mathrm{B} \quad 12\mathrm{B} \quad 7\mathrm{A} \quad 13\mathrm{B} \quad \begin{matrix} 14\mathrm{B} \\ 8\mathrm{A} \end{matrix}$$

and so on *ad infinitum*. The arithmetical definition of A having the same ratio to B which P has to Q, is simply that of A being the same fraction of B which P is of Q: or if

$$A = \frac{l}{k} B \qquad\qquad P = \frac{l}{k} Q$$

Now, since the scale depends entirely on $\frac{l}{k}$, it is the same for both; conversely, if the scale of A and B be the same as that of P and Q, then if $kA = lB$, kP must $= lQ$. Hence the two definitions are synonymous: if one applies, the other does also.

When the multiple scale of A and B is the same as that of P and Q, we have recognised the proportionality of A and B to P and Q. But these scales may differ. The question now is, may they differ in all possible ways, or how far will their manner of differing in one part of the scale affect their manner of differing in others? Am I, to take an instance, at liberty to say, that there may be four magnitudes such that 20 A exceeds 18 B, while 20 P falls short of 18 Q; but that, *for the same magnitudes*, 13 A falls short of 17 B, while 13 P exceeds 17 Q? Such questions as this we proceed to try.

When only two things are possible, which cannot co-exist, each is the complete and only contradiction of the other: the assertion of one is a denial of the other, and *vice versâ*. But when three different things are possible, one only of which can be true, the assertion of one contradicts both of the other two; the denial of one does not establish either of the other two.

The want of a common term, which may simply mean *not less*, that is, either *equal* or *greater*, without specifying which, and so on, causes some confusion in mathematical language. To remind the student that *not less* does not mean greater, but either equal or greater, we shall put such words in italics. Thus, *not less* and *less*, *not greater* and *greater*, are complete contradictions: the denial of one is the assertion of the other.

If A and B be two magnitudes of one kind, and P and Q two others, of the same or another kind, such that

mA is *less* than nB, $\qquad$ mP is *not less* than nQ

then it is impossible that there should be any multiples such that

$m'A$ is *greater* than $n'B$, $\quad$ $m'P$ is *not greater* than $n'Q$

For we find, from the first of each pair,

A is less than $\frac{n}{m}$ B, A is greater than $\frac{n'}{m'}$ B

still more is $\frac{n}{m}$ B greater than $\frac{n'}{m'}$ B or $\frac{n}{m}$ greater than $\frac{n'}{m'}$

But P is *not less* than $\frac{n}{m}$ Q, P is *not greater* than $\frac{n'}{m'}$ Q

Now, all the four combinations of this latter assertion contradict

$\frac{n}{m}$ is greater than $\frac{n'}{m'}$; as follows:

$P = \frac{n}{m}Q$, $P = \frac{n'}{m'}Q$, gives $\frac{n}{m} = \frac{n'}{m'}$

P greater than $\frac{n}{m}$ Q, $P = \frac{n'}{m'}Q$, gives $\frac{n}{m}$ less than $\frac{n'}{m'}$

$P = \frac{n}{m}Q$, P less than $\frac{n'}{m'}$ Q, gives $\frac{n}{m}$ less than $\frac{n'}{m'}$

P greater than $\frac{n}{m}$ Q, P less than $\frac{n'}{m'}$ Q, gives $\frac{n}{m}$ less than $\frac{n'}{m'}$

Hence the two suppositions above cannot be true together: the happening of any one case of either proves every case of the other to be impossible.

If we range all the possible assertions which can be made, we have as follows:

A_3	mA is greater than nB	P_3	mP is greater than nQ
A_2	mA is equal to nB	P_2	mP is equal to nQ
A_1	mA is less than nB	P_1	mP is less than nQ
a_3	$m'A$ is greater than $n'B$	p_3	$m'P$ is greater than $n'Q$
a_2	$m'A$ is equal to $n'B$	p_2	$m'P$ is equal to $n'Q$
a_1	$m'A$ is less than $n'B$	p_1	$m'P$ is less than $n'Q$

Four of these must be true, one out of each triad; and there are 81 ways of taking one of each, so as to put four together. But we shall take the sets A and *a* together, and find what inference we can draw by taking one out of each.

A_3 a_3 proves nothing as to $\frac{n}{m}$ and $\frac{n'}{m'}$. It merely says that $\frac{n}{m}$ B and $\frac{n'}{m'}$ B are both exceeded by A, which may be whether $\frac{n}{m}$ is greater than, equal to, or less than $\frac{n'}{m'}$. The same for P_3 p_3

$A_3\ a_2$ proves $\frac{n'}{m'}$ greater than $\frac{n}{m}$; as does $P_3\ p_2$

$A_3\ a_1$ proves $\frac{n'}{m'}$ greater than $\frac{n}{m}$; as does $P_3\ p_1$

$A_2\ a_3$ proves $\frac{n'}{m'}$ less than $\frac{n}{m}$; as does $P_2\ p_3$

$A_2\ a_2$ proves $\frac{n'}{m'}$ equal to $\frac{n}{m}$; as does $P_2\ p_2$

$A_2\ a_1$ proves $\frac{n'}{m'}$ greater than $\frac{n}{m}$; as does $P_2\ p_1$

$A_1\ a_3$ proves $\frac{n'}{m'}$ less than $\frac{n}{m}$; as does $P_1\ p_3$

$A_1\ a_2$ proves $\frac{n'}{m'}$ less than $\frac{n}{m}$; as does $P_1\ p_2$

$A_1\ a_1$ proves nothing; neither does $P_1\ p_1$

Now, if we put these pairs together, or make pairs of assertions, in the manner already done, we have 81 distinct sets of four assertions, divisible into those which *may* be true together, and those which *cannot* be true together. An inconsequential supposition, such as A_3a_3, may co-exist with any of the rest from the other set Pp; but those which give $\frac{m'}{n'}$ necessarily greater, equal to, or less than $\frac{n}{m}$ in the set Aa, can only co-exist either with the similar ones from the set Pp, or with those which are inconsequential. Thus we have

$A_3\ a_3$	may be true with any marked	Pp				
$A_3\ a_2$	requires either	P_3p_3,	P_3p_2,	P_3p_1,	P_2p_1,	or P_1p_1
$A_3\ a_1$	......	P_3p_3,	P_3p_2,	P_3p_1,	P_2p_1,	.. P_1p_1
$A_2\ a_3$	......	P_3p_3,	P_2p_3,	P_1p_3,	P_1p_2,	.. P_1p_1
$A_2\ a_2$	......	P_3p_3,	P_2p_2,			.. P_1p_1
$A_2\ a_1$	......	P_3p_3,	P_3p_2,	P_3p_1,	P_2p_1,	.. P_1p_1
$A_1\ a_3$	......	P_3p_3,	P_2p_3,	P_1p_3,	P_1p_2,	.. P_1p_1
$A_1\ a_2$	......	P_3p_3,	P_2p_3,	P_1p_3,	P_1p_2,	.. P_1p_1
$A_1\ a_1$	may be true with any marked	Pp				

The remaining thirty cannot be true; but it is unnecessary to specify them, as a simple induction from the preceding will shew how to classify those which may and cannot be true. Attach an idea of magnitude to the phrases *greater, equal,* and *less;* say that "*A is greater than B,*" is higher than "*A is equal to B,*" and this again higher than "*A is less than B.*" We have marked the highest phrases by the highest numbers. Say that in $A_3 a_2$, $A_3 a_1$, &c. (calling A and *a* the *antecedent* clauses of any four marked A, *a*, P, *p*), the antecedents are *descending;* in $A_3 a_3$, $A_2 a_2$, and $A_1 a_1$, *stationary;* and in $A_1 a_2$, $A_1 a_3$, &c. *ascending.* Then all the propositions which imply the co-existence of any two antecedents, and any two consequents of the form A*a* P*p*, may be divided into those which *may* be true, and those which cannot be true, by the two following rules:

Ascending antecedents cannot have descending consequents.

Descending antecedents cannot have ascending consequents.

Precisely the same rules will apply if we take two propositions A P for antecedents, and two others *a p* for consequents; as we may either deduce in the same manner, or by simple inversion. For if A*a* P*p*, with any numerals subscribed, do not contradict either of the preceding rules, neither will the corresponding case of A P *ap* do so, and the contrary. Instances, $A_1 a_2 P_3 p_3$ and $A_1 P_3 a_2 p_3$; $A_2 a_1 P_3 p_2$ and $A_2 P_3 a_1 p_2$, &c.

Let us then take a case of A, B, P, Q, in which we find one ascending assertion relative to *m*A, *n*B, *m*P, *n*Q, for some particular values of *m* and *n;* for instance

$$A_1 P_3 \begin{cases} 3A \text{ is less than } 4B \\ 3P \text{ is greater than } 4Q \end{cases}$$

which, as we have seen, is never contradicted in form by any assertion that can be true of any other multiples. These four quantities are not proportionals: for 3A being less than 4B, and 3P greater than 4Q, P cannot lie in the scale of P and Q in the same place as A in the scale of A and B. But to what more common notion can we assimilate this sort of relation between A, B, P, and Q, namely, that all true assertions of the form (AP) are either ascending or stationary, and never descending? Have we any thing corresponding to this in the arithmetic of commensurable quantities? Let us suppose A and B commensurable, and also P and Q: say that

$$A = \frac{t}{v}B \qquad P = \frac{t'}{v'}Q$$

Then $3\frac{t}{v}$B is less than 4B; $3\frac{t'}{v'}$Q is greater than 4Q;

$\frac{t}{v}$ is less than $\frac{4}{3}$; $\frac{t'}{v'}$ is greater than $\frac{4}{3}$; $\frac{t'}{v'}$ is greater than $\frac{t}{v}$

or A is a less fraction of B than P is of Q; which in arithmetic is also said thus, A bears a less proportion to B than P does to Q, or P bears to Q a greater proportion than A bears to B. Hence we get the following definitions, in which we insert the previous definition of proportion, and the accordance of the whole will be seen.

When all true assertions on $(m\text{A}, n\text{B})$ $(m\text{P}, n\text{Q})$ are either ascending or stationary, and never descending, A is said to have to B a *less* ratio than P to Q; when always stationary, the *same* ratio; when always descending or stationary, and never ascending, a *greater* ratio.

This amounts in fact to the definition given by Euclid, the opening part of whose Fifth Book we shall now make some extracts from, with a few remarks.

DEFINITION III. Ratio is a certain mutual habitude (σχέσις, method of holding or having, mode or kind of existence) of two magnitudes of the same kind, depending upon their quantuplicity (πηλικότης, for which there is no English word; it means relative greatness, and is the substantive which refers to the number of times or parts of times one is in the other).

In this definition, Euclid gives that sort of inexact notion of ratio which defines it in commensurable quantities, and gives some light as to its general meaning. It stands here like the definition of a straight line, "that which lies evenly between its extreme points" prior to the common notion, "two straight lines cannot enclose space," which is the actual subsequent test of straightness. In most of the editions of Euclid we see "Ratio is a mutual habitude of two magnitudes with respect to *quantity*," which makes the definition unmeaning. For quantity and magnitude in our language are very nearly, if not quite, synonymous; or if any distinction can be drawn, it is this: magnitude is the quantity of space in any part of space. But as Euclid is here speaking of magnitude generally (not of space magnitudes only) the words magnitude and quantity are the same.*

* Euclid again uses the word πηλικότης (Book VI. def. 5) in a manner which settles its meaning conclusively. The more advanced reader may consult Wallis, *Opera Mathematica*, v. II. p. 665.

Definition IV. Magnitudes are said to have a ratio to each other which can, being multiplied, exceed "one the other." This means that quantities have a ratio when, any multiples of both being taken, the relation of greater or less exists. It is usually rendered "Two magnitudes are said to have a ratio when the lesser can be multiplied so as to exceed the greater." But the above is literally translated, and the sense here given to ratio makes the next definition consistent. It is a way of expressing that the two magnitudes must be of the same kind, which requires that the notion of greater and less should be applicable to them. That this notion should be applicable to the quantities themselves as well as their multiples, being the necessary and sufficient condition of the possibility of the comparison implied in the next definition, is here assumed * as the distinction of quantities which have a ratio.

Definition V. Magnitudes are said to be in the same ratio the first to the second, and the third to the fourth: when the same multiples of the first and third being taken, and also of the second and fourth, with any multiplication, the first and third (multiples) are greater than the second and fourth together, or equal to them together, or less than them together.

This amounts to our definition of proportion, namely, that the relative multiple scale of A and B is the same as that of P and Q. For, take the same multiples of A and P, namely, mA and mP, and the same multiples of B and Q, namely, nB and nQ. Then, if the relative multiple scales be the same, let mA lie between vB and $(v+1)$B, it follows that mP lies between vQ and $(v+1)$Q. If, then, n be less than v, nB is less than vB, and nQ less than vQ. And mA being greater than vB must be greater than nB, while, for the same reason, mP is simultaneously greater than nQ. In the same way the other parts of the definition V. may be shewn to be included in that of identity of multiple scales. Now, reverse the supposition

* The common version is several times referred to afterwards, and the definition 4 expressly alluded to, in the editions of Euclid. But it must be remembered that the Greek of Euclid contains no references to preceding propositions, these having been supplied by commentators. The reader may, if he can, make Λόγον ἔχειν πρὸς ἄλληλα μεγέθη λέγεται, ἃ δύναται πολλαπλασιαζόμενα ἀλλήλων ὑπερέχειν mean, "Magnitudes are said to have a ratio, when the less can be multiplied so as to exceed the greater."

and assume Euclid's definition. If, then, mA lie between vB and $(v+1)B$, it follows that mA is greater than vB, whence, by the assumption mP is greater than vQ. Similarly, because mA is less than $(v+1)B$, mP is (by that definition) less than $(v+1)Q$. Therefore, mP lies between vQ and $(v+1)Q$, or in this instance, or for any one value of m, the scales are accordant, and the same may be proved in any other case. It follows, then, that the two definitions are mutually inclusive of each other.

The manner in which Euclid arrived at this definition has been matter of inquiry. But any one who will examine the first nine propositions of the *tenth** book, will see that he had precisely the same means of arriving at it as we have used. But, besides this, he might have come by the definition from a common notion of practical mensuration, as follows. Suppose two rods given, one of which is the English yard, the other the French metre, but neither of them subdivided. The only indication which looking at them will offer, is that the metre exceeds the yard apparently by about ten per cent. To get a more exact notion, the obvious plan will be to measure some great distance with both. Suppose 100 yards to be taken off with the yard measure, it will be found that that 100 yards contains about 91 metres and a half, the half being taken by estimation, and we will suppose the eye could not thus err by a quarter of a metre. Then the yard must be ·915 nearly of a metre, and the error upon one yard cannot exceed the hundredth part of the quarter of a metre, or ·0025 of the metre. But the mathematician, to make this process perfectly correct, will suppose distance *ad infinitum*, measured from a point both in yards and metres, or in fact will form what we call the relative multiple scale. He then looks along this scale for a point at which a multiple of a yard, and a multiple of a metre end together. If this happen, and it thus appear that m yards is exactly equal to n metres, the question is settled, for a yard must be $\frac{n}{m}$ of a metre. But it will immediately suggest itself to a mind which is accustomed not to receive assumptions without inquiry, that it may be no two

* There are two English editions of the *whole* of Euclid, and there may be more: that of John Dee (now old and very scarce) and that of J. Williamson, London, 1788, in two thin quarto volumes. The dissertations in the latter are a strange mixture of good and bad, but the text is very literally Euclid, in general.

points ever coincide on the multiple scale. But in this case it is very soon proved that mA may be made as nearly equal to nB as we please, by properly finding m and n; so that a fraction $\frac{n}{m}$ may be found such that A shall be as nearly $\frac{n}{m}B$ as we please. Even admitting that this would do to assign A in terms of B, it leaves us no method of establishing any definite connexion between A considered as a part of B, and P considered as a part of Q.

The word part usually means *arithmetical* part, namely, the result of division into equal parts. Thus $\frac{3}{7}$ is a part of 1 made by dividing 1 into 7 *equal* parts, and taking 3 of them. The phrase of Euclid in the books on number (VII. to X. both inclusive) is that $\frac{1}{7}$ is *part* of 1, $\frac{3}{7}$ is *parts* of 1. And it is easily shewn that, in this use of the word, every quantity is either *part* or *parts* of every other quantity which is commensurable with it. And of two incommensurable quantities, neither is part or parts of the other. But in the original sense of the word part, any less is always part of the greater. This notion of incommensurability, the non-existence of the equation $mA = nB$, for any values of m or n, obliges us to have recourse to a *negative* definition of proportionality, a term which we proceed to explain. Examine the definition of a square, namely, "a plane foursided figure, with four equal sides and one right angle." It is clear that the examination of a finite number of questions will settle whether or no a figure is a square. Has it four sides? are they in the same plane? are the sides equal? is one angle a right angle? *Proof of the affirmative* of these four propositions proves the figure to be a square. Now, examine the number of ways in which a figure can be shewn to be not a square. All propositions are either affirmative or negative; A is B or A is not B. The affirmative can be proved or the negative disproved, with one result only, for both give A is B. But the affirmative can be disproved, or the negative proved, with an infinite number of results; it is done by proving that A is C, or D, or E, &c. &c. *ad infinitum.* Thus there may be an infinite number of ways of shewing that a figure is not a square, but there is only one way of shewing that it is a square. This we call a *positive* definition.

Now examine the definition of parallel lines, " those which are in the same plane, but being produced ever so far do not meet." We are not considering where the lines meet, if they do meet, or distinguishing between lines which meet in one point and in another, but simply dividing all possible pairs of lines into two classes, *parallels* and *intersectors.* Now here it is impossible to prove* the affirmative of the proposition, "A and B are parallels," by means of the definition only, without proving an infinite number of cases. To see this more clearly, remember that every proposition relative to the intersection or non-intersection of straight lines, is an assertion which either includes or excludes every possible couple of points which can be taken, one on each straight line. "Lines intersect" means there is a couple of such points which coincide. "Lines are parallel" means that there is no such couple whatsoever, of all the infinite number which can be taken.

The first proposition in which Euclid *proves* the existence of parallels (the 27th) does not shew that the lines *are* parallels, but that the proposition, "the lines are intersectors," is inconsistent with preceding results. The proposition, "A and B are parallels," though it appears affirmative, yet is in Euclid a negative, for his express definition of parallels does not define what they are, but what they are not, "not intersectors." This we call a *negative* definition.

Now, to examine further Euclid's definition of equal ratios, we must consider his definition of greater and less ratios. They amount to the following. A is said to have to B a greater ratio than P has to Q where there is, among all possible whole numbers m and n, *any one pair* which give mA greater than nB, but mP equal to or less than nQ; or which give mA equal to nB, but mP less than nQ: which give in fact, *in any one case,* what we have called a *descending* assertion. And A is said to have to B a less ratio than P has to Q, when *any one pair* of whole numbers m and n gives mA less than mB, but mP equal to or greater than nQ, or mA equal to nB, but mP greater than nQ: which give in fact, in any one case, what we have called an *ascending* assertion. Here, to a mind the least inquisitive, appears at once a decided objection. Our notions of the terms

* The celebrated axiom of Euclid evades this, and in point of fact amounts to another and a positive definition of parallels, the assumption being that the old definition agrees with it. Or rather we should say, that the first twenty-five propositions of the first book establish a part of the connexion of the definitions, and the axiom assumes the rest.

greater and *less* will never allow us to suppose that any thing, quantity, ratio, or any thing else, can be both greater and less than another quantity, or ratio; and yet, on looking at the definition of Euclid, we see that for any thing which appears to the contrary, one pair of values of m and n may shew that A has a greater ratio to B than P to Q, while another pair may shew that it has a less. The objection is perfectly valid; the only fault to be found is, that it should not have arisen before, when the definitions of the first book were proposed. How is it then known that there can be such a thing as a foursided figure with equal sides and one right angle, or as lines which never meet? The confusion arises from placing the definitions in the form of assertions, before the possibility of the assertions which they imply are proved. The defect may be remedied (we take the square as an instance) in two ways.

1. Write all definitions in the following manner. To define a square, for example, "If it be possible to construct a plane figure having four equal sides and one right angle, let that figure be called a square."

2. Omit the definition of a square, head the 46th proposition of the first book as follows.

"Theorem. On a given straight line, a four-sided figure can be constructed which shall have all its sides equal to the given straight line, and all its angles right angles." Having demonstrated this, add the following definition: Let the figure so constructed be called a square.

We have shewn that all sets of four magnitudes, A and B of one kind, P and Q both of the same kind with the first, or both of one other kind, can be divided into three classes.

1. Those in which simultaneous assertions on mA and nB, and on mP and nQ, are *all* (for all values of m and n) either ascending or stationary.

2. Those in which they are all stationary.

3. Those in which they are all either descending or stationary.

For we have shewn that the only remaining possible case *à priori*, namely, that in which there are both ascending and descending assertions for different values of m and n, is a contradiction amounting in fact to supposing one fraction to be both greater and less than another. And it has been shewn that all the three cases are possible, for commensurable quantities at least. We are now, therefore, in a condition to say, let A and B in the first case be said to have a less ratio to B than P has to Q; in the second, the same ratio; in the third, a less ratio. The only question now is, are these definitions properly negative or positive. It will immediately appear that, out of the three, the first and third can be directly and affirmatively shewn to be true of particular magnitudes, and that the second cannot. By which is meant, that the comparison of individual multiples may, by a single instance, establish the first or third, but that no comparison of individual multiples, however extensive, can establish the second. For the second consists in stationary assertions *ad infinitum*, and the first and third are proved by a single ascending or descending assertion.

As an instance, suppose

A = 951 feet	B = 497 feet	P = 1300 lbs.	Q = 679 lbs.
1902	994	2600	1358
2853	1491	3900	2037
3804	1988	5200	2716
4755	2485	6500	3395

In these first five multiples, there are none but stationary assertions, of twenty five which might be made. Thus

$$\left.\begin{matrix}4755 > 994\\ 6500 > 1358\end{matrix}\right\} \quad \left.\begin{matrix}2853 > 2485\\ 3900 > 3395\end{matrix}\right\} \quad \left.\begin{matrix}951 < 994\\ 1300 < 1358\end{matrix}\right\} \text{ \&c.}$$

but neither of the three definitions is thereby shewn to belong to these four magnitudes. Now, take the first and third 498 times, and the second and fourth 952 times, and we have, going on with the series of multiples,

473598	473144	647400	646408
474549	473641	648700	647087

and here the process may close, for we have 473598 less than 473641, while 647400 is greater than 647087. Consequently, we *have* proved, by comparison, that 951 feet has to 497 feet a less ratio than 1300 lbs. to 679 lbs.

But the case in which neither greater nor less ratio exists can never be established by actual comparison of multiples, except only in the case where the pairs of magnitudes are commensurable. For, remark that the mere circumstance of the relative multiple scale of A and B agreeing with that of P and Q up to any point, is neither proof nor presumption that the two magnitudes given are actually proportional, though, as we shall see, it is certain evidence that they are nearly proportional, if the multiple scales agree for a great number of multiples. Proportion is not established until the similarity of the multiple scales is shewn to continue for ever. Now, though it would not be remarked at first, this insertion of an infinite number of conditions to be fulfilled, is tantamount to a negative definition, if we wish to make the definition specifically speak of one absolute criterion of disproportion or proportion. Disproportion is where there is an ascending or descending assertion somewhere in the comparison of the multiple scales. Proportion is where there is no descending or ascending assertion.

In the case of commensurable quantities the definition is positive, because there is then a single stationary assertion, which, being proved, all the rest are shewn to follow. If A and B be commensurable, let $m\mathrm{A} = n\mathrm{B}$; then if $m\mathrm{P} = n\mathrm{Q}$, there is proportion; if not, there is disproportion. See page 24 for the proof as to the rest of the multiple scales.

We have said, that, when the multiple scales agree for a long period, there is proportion nearly; and it is proved thus: Suppose that the scale of A and B agrees with that of P and Q, up to 10,000 P and 10,000 Q, but that we have disagreement as follows: 9326 A lies between 10,000 B and 10,001 B, whereas 9326 P lies between 10,001 Q and 10,002 Q. Or the scales run thus:

10,000 B	9326 A	10,001 B		10,002 B
10,000 Q		10,001 Q	9326 B	10,002 Q

How much must we alter A to produce absolute proportion? Not more than would be necessary to make 9326 A greater than 10,002 B, or less than would still keep it less than 10,001 B. That is, we must so alter A as to add somewhere between 0 and 2 B to 9326 A, or somewhere between

$$0 \quad \text{and} \quad \frac{2}{9326}\mathrm{B} \quad \text{to} \quad \mathrm{A}$$

Consequently, the addition of a small part of B to A would make an accurate proportion.

We might now proceed to the propositions of the Fifth Book of

Euclid; but there are three difficulties in the way of the student's perfect satisfaction with the definition. 1st, He may have a mysterious idea of incommensurables. 2d, He may not be satisfied of the necessity of departing from arithmetic. 3d, He may find it difficult to imagine how the existence of proportionals can ever be established, with, apparently, an infinite number of conditions of definition to satisfy. We suppose that the gravity of tone which elementary writers adopt, is inconsistent with the statement of a beginner's difficulties, in the words in which he would express them. We shall remove all necessity for preserving such dignity in a case where it may be inconvenient, by a simple supposition. Let *A* be a beginner in the stricter parts of mathematics; that is, a person apt to mix previously acquired notions with the meaning he attaches to definitions which are intended to exclude all but the ideas literally conveyed in the words which are used; much better pleased with the apparent simplicity of an incorrect definition, gained either by omitting what should not be omitted, or by supposing what cannot be supposed, than with the comparatively cumbrous forms which provide for all cases, and distinguish differences which really exist; and, finally, when a doubt exists, rather predisposed against, than in favour of, the necessity of demonstration. Let *B* be another person, who has subjected his mind to that sort of discipline which has a tendency to remove the propensities abovementioned. We can imagine them talking together in this manner:

A.—I have been trying to understand the meaning of incommensurable quantities, and cannot at all make out how it can be that one given line may be no fraction whatsoever of another given line, though both remain fixed, and certain lines ever so little greater or less than the first are fractions of the second.

B.—A little consideration will teach you, that neither in arithmetic nor geometry are we at all concerned with how things *can be*, but only with whether they *are* or not. Do you admit it to be demonstrated that the side and diagonal of a square, for instance, are incommensurable? (*Algebra*, page 98).

A.—I cannot deny the demonstration, but the result is incomprehensible. Does it really prove, that if I were to cut the diagonal of a square into ten equal parts, each of these again into ten equal parts, and so on for ever, I should never, by any number of subdivisions, succeed in placing a point of subdivision exactly upon the point which cuts off a length equal to the side.

B.—I take it for granted you have sufficiently comprehended the definitions of geometry, to be aware that a thin rod of black lead, or a canal of ink, are not geometrical lines; and that the excavations which you perforate by the compasses are not points.

A.—Certainly; I now have no difficulty in imagining mere length intersected by partition marks, which are not themselves lengths.

B.—Then, in the case you proposed, you need not go so far for a difficulty; for your method of subdivision will never succeed in cutting off so simple a fraction as the third part of the diagonal.

A.—Why not?

B.—You see that 9, 99, 999, &c., are all divisible by 3, so that 10, 100, 1000, &c., cannot in any case be divisible by 3, but must leave a remainder. Your method of subdivision can never put together any thing but tenths, hundredths, &c. If possible, suppose one-third to be made up of tenths, a in number, added to hundredths, b in number, added to thousandths, c in number. Then we must have

$$\frac{1}{3} = \frac{a}{10} + \frac{b}{100} + \frac{c}{1000}$$

Clear the second side of fractions, and we have

$$\frac{1000}{3} = a \times 100 + b \times 10 + c$$

or $\frac{1000}{3}$ is a whole number, which is not true. And the same reasoning might be applied to any other case.

A.—This is conclusive enough; but it seems to follow that the third part of a line is incommensurable with the whole.

B.—So it is, as far as the one method of subdividing which you propose is concerned. Let tenths, hundredths, &c., be the *measurers*, and one-third and unity are *incommensurable*. But the word with which we set out implies all the possible subdivisions of halves, thirds, fourths, fifths, &c. &c., to be tried, and all to fail.

A.—But here is an infinite number of ways of subdividing. Can it be possible that no one of them will give a side of a square, when the diagonal is a unit?

B.—In the first place, it would be a sufficient answer to this sort of difficulty to say, that, for any thing you know to the contrary, the number of ways in which you may fail is as infinite as the number of ways in which you may try to succeed. In the second place, there is also an infinite number of ways of subdividing, which will not give one-third. Let your first subdivision be into any number of equal parts, except only 3, 6, 9, 12, &c.; and your second subdivision the same, or any other, with the same exceptions, &c. The same reasoning will prove that you can never get one-third.

A.—But look at the matter in this way. Suppose the halves, the thirds, the fourths, the fifths, &c. &c. of a diagonal laid down upon it *ad infinitum,* so that there is no method of subdividing into aliquot parts, how many soever, but what is done and finished. Would not the whole line be then absolutely filled with subdivision points, and would not one of them cut off a line equal to the side of the square.

B.—You have now changed your use of the word *infinite,* and applied it in the sense of *infinity attained,* not *infinity unattainable.* As long as you used the word to signify succession, which might be carried as far as you pleased, and of which you were not obliged to make an end, the word was rational enough, though likely to be misunderstood; but as it is, you may as well suppose you have got beyond infinite space, at the rate of four miles an hour, and are looking back upon the infinite time which it took you to do it, as imagine that you *have* subdivided a line *ad infinitum.* But if the idea of *infinity attained* be a definite conception of your mind, you meet the difficulty of incommensurable quantities in another form. The definition of the term incommensurable was shaped in accordance with the exact notion, that, subdivide a line as far as you may, you must stop at some finite subdivision; and incommensurable parts of a whole are those which you never exactly separate arithmetically, stop at what finite subdivision you please. But, if you will contend for infinite subdivision attained, and imagine the line thus filled up by points, then it will be necessary to divide all parts of a whole into two classes, those which are cut off by finite subdivision, and those which are not attainable, except by infinite subdivision; the former answering to commensurable, the latter to incommensurable, parts. The difficulty remains then just as before; in other words, why should the side of a *square* be not attainable from its diagonal except by infinite subdivision, when the sides of a rectangle, which are as 3 to 4 (instead of 3 to 3), are attainable by a finite number of subdivisions?

In the next place, you have spoken of a line filled up by points, the infinitude of the number of points being the compensation for each of the points having no length whatsoever; at least, it is not easy to see what else you can mean.

A.—Certainly that is what I mean; and the common expressions of algebra are in accordance with what I say. For, if I cut a line into n equal parts, it is plain that the sum of the n parts makes up the whole, be the number n great or small. But by making n sufficiently great, each of the parts may be made as small as I please; and, therefore, allowing it to be rational to say that P takes place when n is infinite, in all cases in which we may come as near to P as we please, by making n sufficiently great (which is the expressed meaning of *infinite* in algebra), it follows that we may say, that the line is made up of the infinite number of points into which it is cut when divided into an infinite number of equal parts.

B.—I see every thing but the last consequence.

A.—Why, surely, the smaller a line grows, the more nearly does it approximate to a point.

B.—How is that proved?

A.—Suppose two points to approach each other, they continually inclose a length which is less and less, and finally vanishes altogether when the two points come to coincide in one point. So that the smaller the straight line is, the more near is it to its final state—a point.

B.—You have not kept strictly to your own idea (which is a correct one) of the way in which the words *nothing* and *infinite* may be legitimately used. You have supposed a line to be entirely made up of points, each of which has no length whatsoever, because you may compose a line of a very large number of very small lines, each of which, you say, is nearly a point. Let us now consider whether your final supposition is one to which we can approach as near as we please by diminution of a length. Any line, however small, can be divided into other lines by an infinite number of different points; for any line, however small, admits of its halves, its thirds, &c. &c. So that there is a theorem which is not lessened in the numbers it speaks of, or altered in force or meaning, in any the smallest degree, by diminishing the line supposed in it; namely, any line whatsoever admits of as many different points as we please being laid down in it. Now, of your *final length*, or limit of length—the point—this is not true: consequently, you throw away a result at the end, which you cannot throw away as nearly as you please during the process by which you attain that end; nor will the denial of it, near the end, be less in the consequence or amount of the error, than if the rejection were made further from the end. Therefore, in asserting that a diminishing straight line approximates to a point, you have abandoned the condition under which you are allowed to speak of *nothing* or *infinite*.

Again, the nth part of a line taken twice is certainly greater than the simple nth part, however great n may be. Now, what do you suppose two points to be, which are laid side by side without any interval of length between them?

A.—They are, of course, one and the same point.

B.—But in your infinite subdivision, two nth parts must be greater than one nth part, or two of your points must be greater than one; but these two points are the same point, which is therefore twice as great as itself. Such are the consequences to which the supposition of a line made up of points will lead.

A.—I have frequently heard of lines being divided into an infinite number of equal parts.

B.—But you never heard those equal parts called *points*. I can soon shew you that, in the mode of allowing infinity to be spoken of, this fundamental condition is preserved, namely, that no theorem, limitation, number, nor other idea whatsoever, which forms a part of any question, is allowed to be rejected or modified when n is infinite, unless it can be shewn that such rejection or modification may be made with little error when n is great, with less error when n is greater, and so on; finally, with as small an error as we please, by making n sufficiently great. Now, remark the following truths, and the form of speech which accompanies them, when n is supposed infinite.

General Theorem.	Terminal Theorem.
The greater the number of equal parts into which a line is divided, the less *line* is each of the parts: so that an aliquot part of any line, however great, may be made less than any given line, however small.	If a straight line be divided into an infinite number of equal parts, each part is an infinitely small *line*.

General Theorem.	Terminal Theorem.
Any line, however small, may be cut by as many points as we please.	An infinitely small line may be cut by as many points as we please.
No straight line, however small, ceases to be a length terminated by points.	An infinitely small straight line is a length terminated by points.

Now, taking your notion of infinite subdivision attained, it may be shewn that incommensurable parts necessarily follow. For, however far you carry the subdivision, you do not, by means of the subdivision points, lessen the number of points which may be laid down. For each interval defined by the subdivisions contains an infinite number of points. Consequently, if you will suppose the infinite subdivision attained, you cannot do it without supposing an infinite number of points left in the intervals, or an infinite number of incommensurable quantities. This I intend only to shew that the proof of the existence of incommensurable quantities is, upon your own supposition, somewhat better than that of their non-existence. But it would be better to use nothing and infinity as convenient phrases of abbreviation, not as containing definite conceptions which may be employed in demonstration.

A.—I do not see how your objection applies against *nothing;* if we cannot attain infinity by continual augmentation, we can certainly attain nothing by continual diminution.

B.—So it may seem at first, and in truth you are right as to one sort of diminution, that which is implied in the word *subtraction.* From the place in which there is something take away all there is, and you get *nothing* by a legitimate process. But subtraction is the only process which leaves nothing; division, for example, never leaves it. Halve a quantity, take the half of the half, and so on, *ad infinitum:* you will never reduce the result to nothing.

A.—But however clearly you may shew that incommensurable quantities actually exist, as a necessary consequence of our definitions of length, number, &c., I should feel better satisfied if you could give something like an account of the way in which they arise.

B.—If you will consider the way in which number and length are conceived, perhaps the difficulty may be somewhat lessened. Let a point set out from another point, and move uniformly along a straight line until the two are a foot distant from each other. It is clear that every possible length between 0 and one foot will have been in existence at some part or other of the motion. Now, suppose a number of points as great as you please, to set off from the first point together; but, instead of moving in the straight line let them move off in curves, the first coming to the straight line at $\frac{1}{2}$ and 1 of a foot; the second at $\frac{1}{3}$ $\frac{2}{3}$ and 1; the third at $\frac{1}{4}$ $\frac{2}{4}$ $\frac{3}{4}$ and 1 of a foot; and so on, as in this diagram.

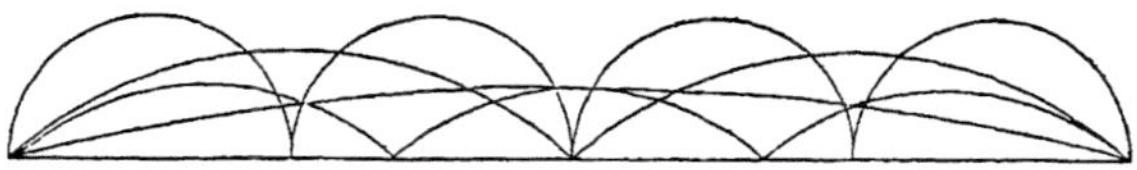

Can you feel sure that these contacts of curves with the line, separated as they must always be from each other by finite intervals, will ever fill up the whole line described by a continuous motion. If not, this figure will always supply presumption in favour of incommensurable parts, which will of course be increased to certainty by the actual proof of their existence. And this should be sufficient to overturn a doubt which after all is derived from confounding the mathematical point with the excavation made by the points of a pair of compasses. The *practical* commensurability of all parts with the whole is a consequence of there being magnitudes of all sorts below the limits of perception of the senses (see page 3).

A.—Granting, then, that there are such things as incommensurable quantities, it is admitted, that though A and B are incommensurable, yet A and B + K may be made commensurable, though it be insisted on that K shall be less than any given quantity, say less than the hundred thousand million millionth of the smallest quantity which the senses could perceive, if they were a hundred thousand million of million of times keener than they are at present. Would it not be sufficient, when incommensurable quantities, A and B, occur, to suppose so slight an alteration made in B as is implied in the above, and reason upon A and B + K so obtained, instead of upon A and B. Surely such a change could never produce any error which would be of any consequence?

B.— Of consequence to what?

A.—To any purpose of life for which mathematics can be made useful.

B.—I am still at a loss.

A.—What process in astronomy, optics, mechanics, engineering, manufactures, or any other part either of physics or the arts of life, would be vitiated by such an alteration, or its consequences, to any extent which could be perceived, were the error multiplied a million fold?

B.—None whatever, that I know of.

A.—What, then, would be the harm of introducing a supposition which would save much trouble, and do no mischief?

B.—I am not aware that I admitted such a supposition would do no mischief, when I said that it would not sensibly vitiate the application of mathematics to what are commonly called the arts of life. I see that your idea of mathematics is very much like that which a shoemaker has of his tools. If they make shoes which keep the weather out, and bring customers, he need not wish them to do more, or inquire further into any use, actual or possible, which they may or might have. The end he proposes to himself is answered, when he has sewed the upper leather firmly to the sole. But whether his art serves any higher purpose—whether the possibility of obtaining conveniencies, and avoiding hardships (which it creates in one respect), excites industry and ingenuity, creates property to equalise the fluctuations of harvests and commerce, and prevent the community from undergoing periodical pests and famines—makes men so dependent on each other that internal war is next to impossible, and external war a grave and serious consideration, &c. &c., are not matters for the thoughts of a working shoemaker; nor will similar considerations ever enter the mind of a working mathematician. You have spoken of the purposes of life; I do not know what the purposes of your life may be, but if among them you count such a discipline of the mind as may always render your perception of the force of an argument properly dependent upon the probability of the premises, and the method by which the inferences are drawn, it will be one of your first wishes to propose to yourself, as a standard and a model, some branch of study in which the first are self-evident, or as evident as any thing can be, and the second indisputable and undisputed. For though you may find no other science which will compete with this in accuracy, yet you will be more likely to infer correctly, when you have seen what you know to be correct inference, than you would have been if you had never, in any case, distinguished between demonstration of certainties and presumptions from probabilities. And still more, will you be qualified to refute, and refuse admission to, that which takes the form of accuracy without the reality. If the mathematical sciences be good as a weapon, they are a hundred fold better as a shield. I have seen many who were visibly little the better for their mathematical studies in what they advanced; but very few indeed who were not made sensibly more cautious in what they received.

A.—But is not my notion adopted in practice by a great part of the mathematical world, particularly on the continent.

B.—It is certainly true, and it is particularly the case with the French, who, though they have done more than any other nation, since the time of Newton, to advance the mathematical sciences, have been by no means anxious to consider them as resting on other evidence than that—not of the senses—but of the limits of the senses. One of their most celebrated elementary writers considers none but arithmetical proportion, and begins his work by shewing either that two straight lines have a common measure, as in page 12, or that the remainder " échappe aux sens par sa petitesse." All his propositions, therefore, *in geometry,* are either true, or so nearly true, that the difference is imperceptible. The phrase we have quoted is an honest and a valuable admission; it shews you, that in the opinion of one of the most useful and extensive elementary writers that ever lived, *arithmetical* proportion makes geometry a science of approximate, not absolute, truth.

A.—I see as much; but cannot the slight shifting of one of the quantities which I proposed be somehow or other corrected, so as to make a strict and useful theory of the proportions of incommensurable quantities?

B.—Yes, and in a very simple way; by adopting the definition of Euclid. This may surprise you, but I will soon shew that the most natural correction of your notion leads direct to the definition of Euclid. Let it be granted that A and B being commensurable, and $m\text{A}=n\text{B}$, proportion between A, B, P, and Q means that $m\text{P}=n\text{Q}$. Now you want, when A and B are *incommensurable,* to be allowed to substitute $\text{B}+\text{K}$ instead of B, where K is excessively small. I suppose you would be perfectly content if it could not be made visible by any microscope. Now I am of a somewhat more abstract turn, and should not like my geometry to be put in peril by the abolition of the excise on glass; which it might be by the allowance of experiments for the improvement of that article, which are now effectually prevented. I cannot admit B + K, where the magnitude I want to reason upon is B. But as the definition of proportion of incommensurables is not yet settled, let us examine this case: A and B being incommensurable, let P and Q be quantities of such a kind that A and B + K are commensurable, and also P and Q + Z, and that the four just named are arithmetically proportionals. Let it be possible, these conditions subsisting, to make K and Z as small as we please: not as small as this, that, or the other small quantity, but smaller than any whatsoever which may be named, being still *some* quantities. You wish to substitute B + K and Q + Z for B and Q: I prefer to use the conditions laid down to ascertain how B and Q themselves stand related to A and P. Let us suppose we name two small magnitudes, K′ and Z′, of the same kind as A and P, or B and Q, or K and Z, which we are at liberty to make as small as we please. We can then find K and Z less than K′ and Z′, and such that A, B + K, P, Q + Z are proportional. Suppose A and B + K commensurable, and let

$$mA = n(B + K) \quad \text{whence} \quad mP = n(Q + Z)$$

whence it is easily proved, as in page 24, that the relative scale of multiples of A and B + K is the same as that of P and Q + Z. I say it follows, that the relative scale of A and B is the same as that of P and Q; for, if not, the two latter scales must differ somewhere. Let it be that vA is greater than wB, but vP less than wQ. Then, since vA is greater than wB, let K be taken so small (it may be as small as we please) that vA shall also exceed $w(B + K)$, whence, by the proportion assumed in the hypothesis, vP exceeds $w(Q + Z)$, while, by the hypothesis we are trying, vP is less than wQ. This is a contradiction, for vP cannot exceed $wQ + wZ$ and fall short of wQ at the same time. In the same way, any other case may be treated; and it follows that our suppositions, if K may be as small as we please, amount to an hypothesis from which Euclid's definition follows. If, in the above, we suppose vA less than wB, while vP is greater than wQ, we see that vA also falls short of $v(B + K)$, whence, by the proportion, vP falls short of v(Q + Z), which cannot be if it exceed vQ.

A.—But is not this deduction, namely, Euclid's definition, more cumbrous than the form from which it has just been deduced?

B.—How so?

A.—Does it not involve an infinite number of considerations, extending the whole length of the multiple scales?

B.—And does not your definition do the same thing, unless you stop somewhere with the values of K and Z? Is it not necessary, if we would not be merely microscopically correct, but absolutely correct, to suppose that K and Z may be diminished and diminished *ad infinitum?* And what difference is there, as to the number of considerations in question, between two magnitudes which are to diminish without limit, and a set of increasing multiples of two given magnitudes?

A.—But Euclid's definition seems to wander such a way from the quantities in question, while the other remains close to them, and we never seem to quit them, except for something very near to them. The actual application of the definition I prefer will require nothing but the division of all magnitudes into aliquot parts.

B.—Your objection amounts to this; that you feel the fractions of a quantity to be more closely connected in your mind with the quantity itself than its multiples. This may be the case; and, if so, it is some reason for preferring the form to which you seem most inclined. But there may be a stronger reason for preferring the other; and, undoubtedly, as long as difficulties exist, every system of science must be a balance of inconveniences. But Euclid is, of all men who ever wrote, the one who has a reason for the course which he takes, where there are two or more. I suppose you cannot but admit that it is better to found a definition *in geometry* upon the result of something which can actually be done, *by the means of geometry*, than upon something which can only be conceived or imagined to be done, with what certainty soever; for instance, you would not wish to be obliged to use other means than the straight line and circle, or to suppose an object gained without using any means at all?

A.—Certainly not. That there shall be no assumption of mechanical power beyond that of drawing a straight line or circle, is the foundation of pure geometry.

B.—Then the question is settled in favour of Euclid's definition;

for, without either assuming more mechanical means, or making a gratuitous assumption, no angle, nor arc, nor sector of a circle, can be divided into 3, or 6, or 9, &c. parts, unless it be a right angle, or a given half, fourth, eighth, &c. of a right angle. There are some other exceptions; but, generally, to cut any angle into three equal parts is a geometrical impossibility, and certain algebraical considerations furnish the highest presumption that it will always remain so.

A.—But this difficulty is still left: how are we ever to shew that there are such things as proportional quantities?

B.—We can do this so easily, that the greatest stumbling-block of the process lies in its being so easy and perceptible, that a beginner does not very well see where lies the knowledge he has gained, unless he has paid profitable attention to the definition of proportion. From the first book of Euclid it is evident that a rectangle is doubled by doubling the base, trebled by trebling it, and so on; and also, that of two rectangles between the same parallels, the greater base belongs to the greater, and the lesser base to the lesser. Now, let B and B′ represent two bases, and R and R′ the rectangles upon them, the altitudes, or distances of the parallels, being the same. If then we take the first base m times, giving mB, the rectangle upon that base is mR: if we take the second base n times, giving nB′, the rectangle upon that base is nR′, the parallels always remaining the same. Hence it follows, that mB and nB′ are bases to the rectangles mR and nR′ between the same parallels; accordingly, therefore, as mB is greater than, equal to, or less than nB′, so is mR greater than, equal to, or less than nR′: and this being true for all values of m and n, it follows that B has to B′ the ratio of R to R′, or the bases of rectangles between the same parallels, and the rectangles themselves are proportionals.

A.—Am I to understand then that there are difficulties in the way of considering magnitude in general, which are not found in arithmetic, or the science of abstract number?

B.—Quite the reverse: the difficulty arises from the deficiences of arithmetic itself, and from their being ratios which the ratio of number to number cannot represent.

A.—But how is that? arithmetic always seemed clear of such difficulties as we have been considering.

B.—And so would this subject, if the disposition to be satisfied with what is in the book, which is part and parcel of almost every beginner had been permitted to rest quietly upon a theory of commensurable ratios. But did you never, in arithmetic, hear of the creation of a nonexistent number or fraction, in spite of there being no such thing, by agreeing that there should be such a thing, and drawing a picture to represent it?

A.—I do not understand the jest; but I suppose you allude to algebra, and to quantities less than nothing?

B.—Not at all; I am speaking of pure arithmetic. To me, -2 is a much easier symbol, or picture, than $\sqrt{2}$; and even the difficulties of $\sqrt{-2}$ lie as much in the $\sqrt{\ }$ as in the $-$.

A.—But I do not understand what you mean by saying that $\sqrt{2}$ does not exist; it is the square root of 2, and multiplied by itself it gives 2. You may find it as nearly as you please.

B.—If it be the object of arithmetic, commonly so called, it is either a whole number or a fraction. Which of these is it?

A.—It is a fraction; 1·4142136, very nearly.

B.—I did not ask you what it is *very nearly*, but what it is?

A.—It cannot be given exactly, but we all know there is such a thing as the square root of 2.

B.—If the objects of arithmetic were numbers, fractions, and things, and the latter term had a definition, I might admit what you say. And in *concrete* arithmetic, where 1 is a *thing*, a foot, a pound, or an acre, I admit that there is such a thing as $\sqrt{2}$. But that thing is not attainable arithmetically by taking any aliquot part of the thing 1, and repeating it any number of times. In abstract arithmetic the square root of 2 is an impossibility; and having no existence, I do not see how one fraction can be said to be nearer to it than another, except in this sense, that $2+z$ may be made to have a square root where z may be less than any fraction we name. The independent existence of $\sqrt{2}$ is an algebraical consideration of some difficulty; that is, belongs to the science which has relations of symbols, under prescribed definitions, for its object, without reference to their numerical interpretation. The difficulties of $\sqrt{2}$ are precisely those of incommensurable magnitudes; in fact $\sqrt{2}$ is the diagonal of a square whose side is 1. But it is to algebra that difficulties of this kind should be referred. The student, if he use $\sqrt{2}$ in pure arithmetic, must expressly understand it as a fraction whose square is nearly 2, and must consider this part of arithmetic (without algebra as a science of approximation, unless geometry, or some other science of concrete quantity, be supposed to lend its aid.

A.—But I cannot divest myself of the idea that $\sqrt{2}$, $\sqrt{3}$, and $\sqrt{6}$ are really fractions, and that the product of the two first gives the last. I suppose, in some sense or other, you admit this proposition?

B.—Certainly. If $\sqrt{2+x}$ and $\sqrt{3+y}$ and $\sqrt{6+z}$ be made to exist, by giving proper values to x, y, and z, which may all be as small as I please, and if, moreover, x, y, and z be so related that $z=3x+2y+xy$, which condition does not interfere with the last; I can then admit that

$$\sqrt{2+x} \times \sqrt{3+y} = \sqrt{6+z}$$

But I do not allow myself to suppose that (understanding by multiplication the taking of one number or fraction as many times or parts of times as there are units or fractions of a unit in another), there can be such a truth as that

$\sqrt{2}$ (neither number nor fraction) *multiplied* by $\sqrt{3}$ (do. do.) $=$ $\sqrt{6}$ (do. do.)

But this is beyond our subject, except so far as it shews that the difficulty lies more in arithmetical than in geometrical considerations.

A.—Might we not then dispense with arithmetic altogether, and make a definition corresponding to proportion for geometry?

B.—Yes; but the difficulty would appear in another shape, of the very same substance. Let four lines be called proportional when, being straightened without alteration of length, if necessary, the rectangle made by the first and fourth is equal to that made by the second and third. Let areas be proportional when, being converted into rectangles with a common altitude, their bases are proportional. Let angles be proportional, when they are angles at the centre of proportional arcs of the same circle. But here would immediately arise this difficulty,—to make a straight line equal to a given arc of a circle; which is out of the power of the geometry of straight lines and circles.

A.—Is not the *reductio ad absurdum* (which is very much used in the establishment of the theory of proportion) rather a suspected method. I have heard it called *indirect* demonstration; and it is frequently stated as a defective method, not to be used if it can possibly be avoided.

B.—The complaints against this method of demonstration have become much more frequent, if not entirely made their appearance, since the time when logic was a necessary part of a liberal education, as it once was, and as I hope it will be again. I have sometimes wondered whether this argument would have been considered objectionable if it had been reduced to the form "A is B, B is C, therefore A is C;" as follows: "Every contradiction of P is a contradiction of the proposition that the whole is greater than its part; but every contradiction of this proposition is false: therefore every contradiction of P is false; or P is true." The *reductio ad absurdum* is as conclusive, and may be made as intelligible, as any other argument. And if any argument be good in proportion to the effect upon the mind, where is the affirmative proposition, in geometry or not, which the mind seizes as readily as it recoils from an absolute contradiction in terms? Where is the likeness or resemblance between things which are alike, that is so forcible as the unlikeness or want of resemblance of two ideas which palpably contradict, such as black is white?

A.—Is there then no advantage in the direct over the indirect demonstrations?

B.—D'Alembert has said that the former are to be preferred "parce qu'elles éclairent en même-temps qu'elles convainquent," which is a good description of the difference. But even this must be taken with some allowance, for there are many indirect demonstrations which are highly instructive.

Recapitulation. By the *ratio* of A to B, we mean (without any further specification at present) a relation between the magnitudes of A and B, determined by the manner in which the multiples of A are distributed, if each be written between the nearest multiples of B in magnitude. That is, if B, 2 B, 3 B, &c., be formed, and A, 2 A, 3 A, &c., and if A lie between B and 2 B, 2 A between 3 B and 4 B, and so on, the relative scale

B, A, 2 B, 3 B, 2 A, 4 B, &c.

is to be the sole determining element of the ratio, so that there is to be nothing but the order of this scale on which the ratio depends. And if P and Q be two other magnitudes with the same order in their scale, P compared with A, and Q with B, then A and B are to be said to have the same ratio as P and Q. But if any multiple of A precede among the multiples of B the place which the corresponding multiple of P occupies among the multiples of Q, then A is to be said to have to B a less ratio than P has to Q. But if a multiple of A come later in the series of multiples of B than its corresponding multiple of P in the series of multiples of Q, then A is said to have to B a greater ratio than P has to Q. It is plain that the ratio of A to B must be greater than, equal to, or less than that of P to Q, and also, that in saying A is to B as P to Q, we also say that B is to A as Q to P.

[We must remind the student that we have now nothing to do with the reasons of this definition, or the accordance of its parts with each other, or with any notion of ratio more than is contained in it. We are merely now concerned to know what follows from this definition. The numbering of the following propositions is that in Euclid.]

When A has to B the same ratio as P to Q, the four are said to be proportionals, and are written thus:

$$A : B :: P : Q$$

which is read A is to B as P is to Q.

IV. If $A : B :: P : Q$
Then $mA : mB :: nP : nQ$ } m and n being any whole numbers.

This we know when we see that any quantities being arranged in order of magnitude, so will be their multiples. If the scales be

$$\begin{array}{lllll} B & A & 2B & 3B & \ldots\ldots \\ Q & P & 2Q & 3Q & \ldots\ldots \end{array}$$

the following scales

$$\begin{array}{lllll} mB & mA & 2mB & 3mB & \ldots\ldots \\ nQ & nP & 2nQ & 3nQ & \ldots\ldots \end{array}$$

will also be arranged according to magnitude. Whence the proposition.

VII. If A, B, C, be three *homogeneous* magnitudes (all lines, or all weights, &c.) and if $A = B$; then

$$A : C :: B : C$$

and

$$C : A :: C : B;$$

for the scales must evidently be identical.

VIII. A+M has a greater ratio to B than A has to B, and B has a less ratio to A+M than B has to A. Let M be multiplied so many times that it exceeds B; say $m\text{M} = \text{B} + \text{K}$: then

$$m(\text{A} + \text{M}) = m\text{A} + \text{B} + \text{K}$$

Let $m\text{A}$ lie between $v\text{B}$ and $(v+1)\text{B}$; then $m(\text{A}+\text{M})$ lies between $v\text{B}+\text{B}+\text{K}$ and $(v+1)\text{B}+\text{B}+\text{K}$, and certainly beyond $(v+1)\text{B}$. Consequently, in the scales of A+M and B, and A and B, a multiple of A + M is found to be in a higher place among the multiples of B than the same multiple of A among the multiples of B. Whence, by definition, A+M has to B a greater ratio than A to B. The second part of the proposition is but another way of stating the first, as appears from definition. Thus we may also say that A has to B a less ratio than A + M has to B.

IX. If $\quad \text{A} : \text{C} :: \text{B} : \text{C} \quad$ then $\quad \text{A} = \text{B}$

or if $\quad \text{C} : \text{A} :: \text{C} : \text{B} \quad$ then $\quad \text{A} = \text{B}$

For (VIII.), if A be greater than B, A has to C a greater ratio than B to C, which is not true. If A be less than C, A has to C a less ratio than B to C, which is not true: therefore A=B. The same reasoning proves the second case.

X. If A have to C a greater ratio than B has to C, then A is greater than B. For if A were equal to B, then these ratios would be the same: if A were less than B (VIII.), then would A have to C a less ratio than B has to C. Therefore, A is greater than B. Similarly, if A have to C a less ratio than B has to C, A is less than B. And if C have a greater ratio to A than to B, A is less than B; if C have a less ratio to A than to B, A is greater than B.

XI. If the ratios of C to D and of E to F, be severally the same as that of A to B, then C has to D the same ratio as E to F. This answers to a case of the general axiom, that two things which are perfectly like to a third in any respect, are perfectly like each other in that respect. The multiples of C are distributed among those of D in the same manner as those of A among those of B, as are those of E among those of F. Therefore, the multiples of C are distributed among those of D as are those of E among those of F. Whence the proposition.

XII. If A be to B as C to D, and as E to F, then as A is to B so is A+C+E to B+C+F.

For mA lying between nB and $(n+1)B$, then mC lies between nD and $(n+1)D$, and mE between nF and $(n+1)F$, and, consequently, $mA+mC+mE$, or $m(A+C+E)$ between $n(B+D+F)$ and $(n+1)(B+D+F)$. Whence the proposition.

XIII. If A have to B the same ratio as C has to D, but C to D a greater ratio than E to F, then A has to B a greater ratio than E to F.

This is one of a class of propositions which come under this general theorem: for any ratio, an equal ratio may be substituted, and all consequences of the first ratio are consequences of the second. This, which seems very evident, may appear so upon mistaken evidence. Ratios, as far as we have yet gone, are not quantities, but expressions of that relation between quantities upon which the order of magnitude of their multiples depends. For quantity, we may substitute other quantity equal to the first in magnitude wherever the relation is one which depends only on quantity; we may not substitute a triangle of the same area instead of a square, except there be question of nothing but superficial magnitude, or area. Ratio, again, is to us at present the order of the multiples, so that if A and B have their multiples arranged among each other in a given order, if P and Q have the same, we may say that whatever is true of the order of multiples of A and B, is also true of the order of P and Q; whatever connexion the order of multiples of A and B establishes between A and B and other magnitudes, the same connexion exists between P and Q and those other magnitudes, because the accident of A and B, which is the sole connexion between them and the consequence inferred, is also an accident of P and Q. The necessity for going over such considerations, arises from its never being allowed to be taken for granted that a mathematician has studied logic. Hence Euclid* is frequently obliged to reiterate the same assertions in different forms. To take the proof of the present proposition; to say that C has to D a greater ratio than E to F, is to say that mC can be found greater than

* Euclid was a contemporary of Aristotle, as is generally supposed, and may, therefore, never have seen the science of the latter. It is free to us to suppose that if he had, he would have distinguished between a purely logical and a geometrical consequence: that is, would not have reiterated the same proposition in different forms; or, if you please, different cases of the same verbal truth as if they were distinct truths: and we will suppose so accordingly.

nD, while mE is equal to or less than nF. But to say that A has to B the ratio of C to D, is to say that whenever mC is greater than nD, mA is greater than nB. Therefore, to say that the ratios of A and B and C and D are the same, but the latter greater than that of E to F, is to say that mA may be greater than nB, while mE is equal to or less than nF; or that A has to B a greater ratio than E has to F.

Now, let the student compare this with the following proposition. A and B are greens of exactly the same shade: but B is a darker green than C, therefore, A is a darker green than C. Would it be unnecessary to prove this? then it is equally unnecessary to prove the preceding. But we will prove this in the same manner as we prove the preceding. Let there be a test of greenness, which decides between two greens (there is a test of comparison of ratios in Euclid), and apply the test to B and C. The result is, of course, that B is the darker. But A being by hypothesis exactly the same as B, the testing operation would be self contradictory if it did not exhibit, when applied to A and C, the very same intermediate process by which we were able to compare B and C, with the same result. If the above be unnecessary, then the demonstration of Euclid's proposition is unnecessary.

The fact is, that there are in geometry two distinct sorts of demonstration, the first of which is only a portion of the second. The first is the verbal treatment of the terms of an hypothesis, and the developement of all assertions which are necessarily included in the terms of the proposition, without drawing upon any other axioms or theorems for evidence. It is the purely logical process, by which we make two assertions put together shew their joint meaning, and express what, without deduction, they only imply. Thus, from "Every A is B," and "no B is C," we make it evident that in these assertions is necessarily contained a third, that "no A is C." Thus it has been shewn that we cannot allow simultaneous existence to the two propositions, "A is to B as C is to D," and "C is to D more than E is to F," without almost expressing, and certainly implying, by the mere meaning of our terms, this third proposition, that "A is to B more than E is to F."

The second process is that in which the demonstration, besides the purely logical process of extracting implied meanings out of the expressions of the hypothesis, appeals to propositions which are not in the hypothesis, and which, for any thing the hypotheses tell us to the contrary, may or may not be true. Of course—not logic, but—reason requires that these propositions should have been previously proved, or assumed on their own evidence expressly. Let us take the following proposition, "The sum of the circles described upon the two sides of a right-angled triangle is equal to the circle described upon the hypothenuse." Now, take every notion implied in this hypothesis, "Let there be a right-angled triangle, and let circles be described on its three sides." The united faculties of man never proved that the sum of the circles on the sides was equal to the circle on the hypothenuse, without assuming with Euclid, to the effect that only one parallel can be drawn through a point to a given right line; with Archimedes, to the effect that the chord of a curve is shorter than its arc, &c. &c.; and various consequences. But are any of these propositions necessary to our complete definition of a right angle, a triangle, or a circle? If not, we have a broad and easily recognised distinction between the first and second method of demonstration; the first, an operation of logic, or deduction from the premises of the hypothesis; the second, introducing premises from without.

There are two classes of reasoners whose ideas we recommend the student closely to examine, before he finally decides: 1. Geometrical writers in general, who pay no attention to the methods which they are using, but let the first book and the fifth book of Euclid contain no difference by which it may be remarked that the processes contained in the two are different acts of mind. Did they ever think that geometry could be made the engine by which the student could examine certain operations of his own faculties, or did they only imagine that it was a method of making very sure that squares, circles, &c. had such and such properties? 2. The class of metaphysical writers, who express themselves to the effect that all mathematical propositions are contained in the definitions and axioms, in a sense in which other results of reasoning are not. Put them to the proof of this assertion as to geometry, and then as to arithmetic.

The whole of the process in the fifth book is purely logical, that is, the whole of the results *are* virtually contained in the definitions, in the manner and sense in which metaphysicians (certain of them) imagine all the results of mathematics to be contained in their definitions and hypotheses. No assumption is made to determine the truth of any consequence of this definition, which takes for granted more about number or magnitude than is necessary to understand the definition itself. The latter being once understood, its results are deduced by inspection—of itself only, without the necessity of looking at any thing else. Hence, a great distinction between the fifth and the preceding books presents itself. The first four are a series of propositions, resting on different fundamental assumptions; that is, about different kinds of magnitudes. The fifth is a definition and its developement; and if the analogy by which names have been given in the preceding books had been attended to, the propositions of that book would have been called *corollaries of the definition.*

XIV. If A be to B as C is to D, all four being of the same kind, then if A be greater than C, B is greater than D; if equal, equal, and if less, less.

A must either be $>$ $=$ or $<$ C. Let A be greater than C; then mA is greater than mC. Let mA lie between nB and $(n+1)$B; then will mC lie between nD and $(n+1)$D. But because A exceeds C, 2A exceeds 2C by twice as much, &c., and mA exceeds mC by m times as much; or mA may be made to exceed mC by a quantity greater than any one named, say greater than B and D together. Then the order of magnitude of the four multiples mC $(n+1)$D, nB, mA must be as written: for $(n+1)$D does not exceed mC by so much as D, and nB does not fall short of mA by as much as B, while mA exceeds mC by more than B and D put together. Therefore, nB is greater than $(n+1)$D, and still more than nD. That is, B is greater than D.

Let A be equal to C. If B exceed D at all, mB may be made to exceed mD by more than D, or mB may be made, from and after some value of m, greater than $(m+1)$D. That is, the order of magnitude may be made

$$m\text{D} \qquad (m+1)\text{D} \qquad\qquad m\text{B} \qquad (m+1)\text{B}$$

Having gone so far on the scales that this order becomes permanent, go on till a multiple of C (kC) falls between the two first. Then, by the definition, kA falls between the two last, which is absurd; for, because A$=$C, kA$=k$C; therefore, B does not exceed D. In the same way it may be shewn that B does not fall short of D. Therefore, B$=$D.

The remaining case (A less than C) may be proved like the first.

XV. A is to B as mA is to mB

The scale of multiples of A and B is nowhere altered in the order ot magnitude by multiplying every term by m. If pA lie between qB and $(q+1)$B, (pm)A which is $p(m\text{A})$ lies between $q(m\text{B})$ and $(q+1)\ (m\text{B})$.

XVI. If A be to B as C is to D

and if all four be of the same kind,

Then A is to C as B is to D.

(IV.) mA is to mB as nC is to nD

(XIV.) If mA be greater than nC, mB is greater than nD, if equal, equal; if less, less. Therefore, A is to C as B to D.

XVII. If A + B be to B as C + D to D, then A is to B as C is to D. If mA lie between nB and $(n+1)$B, it follows that $m\text{A}+m\text{B}$, or $m(\text{A}+\text{B})$ lies between $(m+n)$B and $(m+n+1)$B. Then, by the proportion, $m(\text{C}+\text{D})$ lies between $(m+n)$D and $(m+n+1)$D, or $m\text{C}+m\text{D}$ lies between $m\text{D}+n\text{D}$ and $m\text{D}+(n+1)\text{D}$, or mC lies between nD and $(n+1)$D. Therefore, the scales of A and B, and of C and D, are the same; whence the proposition.

XVIII. If A be to B as C is to D, then A + B is to B as C + D is to D. A proof of exactly the same kind as the last should be given by the student.

XIX. If A : B :: C : D, C and D being less than A and B, then A : B :: A — C : B — D. For the hypothesis gives A to C as B to D, and A is C + (A — C), and B is D + (B — D), whence,

C + (A — C) is to C as D + (B — D) is to D

(XVII.) A — C is to C as B — D is to D

(XVI.) A — C is to B — D as C to D, or as A to B

XX. If A be to B as D to E

and B to C as E to F

Then A is $\left\{\begin{array}{l}\text{greater than}\\ \text{equal to}\\ \text{less than}\end{array}\right\}$ C, when B is $\left\{\begin{array}{l}\text{greater than}\\ \text{equal to}\\ \text{less than}\end{array}\right\}$ F

Let A be greater than C; then A is to B more than C is to B; but A is to B as D to E, and C to B as F to E; therefore, D is to E more than F is to E, or D is greater than F. In a similar way the other cases may be proved.

Hence it follows, that A is to C as D to F. For,

(VI.)	mA	is to	mB	as	mD	to	mE
	nB	is to	nC	as	nE	to	nF

therefore, mA is $> =$ or $< nC$ when mD is $> =$ or $< nF$

whence, A is to C as D to F.

XXI. If of the magnitudes

$$\begin{matrix} A & B & C \\ D & E & F \end{matrix} \quad \text{we have} \quad \begin{matrix} A : B :: E : F \\ B : C :: D : E \end{matrix}$$

Then $A > =$ or $< C$ when $D > =$ or $< F$

Let A be greater than C; then A is to B more than C is to B: as before E is to F more than E is to D, or D is greater than F. Similarly for the other cases.

XXII. If there be any number of magnitudes,

$$\begin{matrix} A & B & C & D & \ldots\ldots\ldots \\ P & Q & R & S & \ldots\ldots\ldots \end{matrix}$$

and if any two *adjoining* be proportional to the two under or above them, then any two whatsoever are proportional to the two under or above them. For, since (XX.)

$$\left.\begin{matrix} A : B :: P : Q \\ B : C :: Q : R \end{matrix}\right\} \begin{matrix} \text{Therefore, } A : C :: P : R \\ \text{But, } \quad C : D :: R : S \end{matrix}\left.\vphantom{\begin{matrix} A \\ B \end{matrix}}\right\} \text{Therefore, } A : D :: P : S, \text{ \&c.}$$

XXIII. In the hypothesis of (XXI.), by proof as before in (XX.), A is to C as D to F.

XXIV. If A be to B as C to D
and E be to B as F to D } then A + E is to B as C + F to D

For, A : B :: C : D
and B : E :: D : F } whence, A : E :: C : F

(XVIII.) A + E : E :: C + F : F

But, E : B :: F : D

Therefore, A + E : B :: C + F : D

XXV. If A : B :: C : D, all being of the same kind, the sum of the greatest and least is greater than that of the other two. First, which are the greatest and least? If A be the greatest, then C is greater than D; and because A : C :: B : D, B is greater than D;

therefore, D is the least. Now, prove that if B be the greatest, C is the least; and that, by inverting the proportion, if necessary, it may always be written with the greatest term first, and the least last.

When A is the greatest, since $A-B : A :: C-D : D$, $A-B$ is greater than $C-D$; therefore, $(A-B)+B+D$ is greater than $(C-D)+B+D$, or $A+D$ is greater than $C+B$.

If there be a given ratio, that of A to B, and another magnitude P, there must be a fourth magnitude Q, of the same kind as P, such that A is to B as P to Q, or Q to P as B to A.

Firstly; Q may certainly be taken so small that (mB being greater than nA) mQ shall be less than nP. Find m and n to satisfy the first conditions, and let K satisfy the second. Then K is to P less than B is to A. Now (mB being less than nA), Q may be taken so that mQ shall be greater than nP. Find m and n to satisfy the first, and let L satisfy the second. Then, L is to P in a greater ratio than A to B. And it is immediately shewn that every magnitude less than K is to P less than B to A, and every magnitude greater than L is to P more than B to A. Whence, it is between K and L that the fourth proportional Q is found, if any where. There cannot be more than one such value of Q; for, if there be two different magnitudes V and W, since, then, by taking m sufficiently great, we may make mV and mW differ by more than P, it is impossible that both mV and mW can lie between the same consecutive multiples of P, as those of B which contain between them mA. And the above also evidently shews, that if we suppose a magnitude Q, changing its value from K to L, it cannot during its increase become of the same kind as L, namely, more to P than B is to A, and then again become of the same kind as K. For, whatever magnitude has this property of L, every greater one has the same. There is then only one point between K and L at which this change takes place, and we have, therefore, this alternative: EITHER G (between K and L) is less to P than B is to A, and *every* magnitude greater than G is more; OR, some magnitude G between K and L is the same to P as B to A, and is the intermediate limit lying above all those which are less to P, and below all those which are more. By disproving the first alternative, we prove our proposition. If possible, let G be less to P than A to B, G+V more, however small V may be.

Then may mG be made less than nP (mA being greater than nB), while $m(G+V)$ is greater than mP. For the *ascending* assertion must be converted at least into a *stationary* one. Let mG fall short of mP by Z; then V may be taken so small that mV shall not be so great as Z, or $mG+mV$ not so great as $mG+Z$, that is, not so great as mP. But the first clause of the alternative supposes that $m(G+V)$ must be greater than mP, how small soever V may be; therefore this clause cannot be true, or the second must be true.

This fourth proportional to A, B, and P, then, must exist; but whether it can be expressed by the notation, or determined by the means of any science, is another question. It can be expressed in arithmetic when A and B are commensurable: it can be found in geometry (by the straight line and circle) when A and B are lines or rectilinear areas. But if they be angles, arcs of circles, solids, &c. it cannot be assigned by the straight line and circle, except in particular cases.

Let us suppose the ratio of A to B given, that is, not A and B themselves, but only the answer to this question for all values of m, "Between what consecutive multiples of B lies mA?" Suppose also the ratio of B to C given; how are we to find the ratio of A to C, or can it be found at all? that is, is it given or determined by the two preceding ratios. Take any magnitude P, and determine Q so that P is to Q as A to B, and then determine R so that Q is to R as B to C. Then the ratio of P to R (page 59) is that of A to C; not that P is A or R is C (for they may even be magnitudes of different kinds), but P is to R as A is to C.

The process by which the ratio of A to C is found by means of those of A to B and B to C, is called by Euclid *composition* of these ratios; or the ratio of A to C is compounded of the ratios of A to B and B to C. What, then, ought to be meant by the ratio compounded of the ratios of A to B and X to Y. Our guide in the assimilation of processes, and the extension of names, is always the following axiom.

Let names be so given, that the substitution of one magnitude for another equal magnitude shall not change the name of the process; and, generally, that the same operations (in name) performed upon equal magnitudes, shall produce the same result.

Let X be to Y as B to N, where N is a fourth proportional to be determined. Then the ratio of A to N is that compounded of A to B and B to N, and is what must be meant by that compounded of

A to B, and X to Y. It is proved in Prop. 20, that ratios compounded of equal ratios are equal ratios.

Again, to find the ratio compounded of the ratios of A to B, C to D, and E to F; let the process by which the ratio of A to D is derived from those of A to B, B to C, and C to D, still be called composition. Then take B to M as C to D, and M to N as E to F: the ratio of A to N is that compounded of the three ratios.

In the beginning of this work, we deduced the necessity for considering incommensurables in some such manner as that of Euclid, from the notion which, as applied to commensurables, admits of a definite representation, derived from the idea of proportion. But the method of the fifth book is different. It is there implied, that whereever two magnitudes exist, their joint existence gives rise to a third magnitude, called their ratio, of which magnitude no conception is given except what is contained in certain directions how to apply the terms equal, greater, and less, to two of the kind. On this the natural question is, what sort of magnitude is this, and how do we know that there is any magnitude whatsoever which admits of this apparently arbitrary exposition of definitions? This question is very much to the point, and the want of an answer at the outset is a main cause of the difficulty of the Fifth Book. The answer implied in the work of Euclid is this: Let us first consider what will follow if there be such things as ratios, or magnitudes to which these definitions of equal, greater, and less apply; we shall then shew (in the Sixth Book) that there are different pairs of magnitudes, of which it may be said that they have ratios, and we shall never have occasion to inquire what ratio is.

We may take a case parallel to the preceding from the First Book. The notion of a straight line suggests nothing but length; that of two straight lines which meet, suggests a relation, which we may conceive stated in this way. If A, B, C, and D, be straight lines, of which A and B, and C and D, meet; let A and B be said to make the same angle as C and D, when, if A be applied to C, and B and D fall on the same side, B and D also coincide: but let A be said to make a greater angle with B than C with D, when, in a similar case, B falls outside of C and D, &c. To this it would be answered, that the preceding definitions are a circuitous way of saying that the angle made by two lines is their opening or inclination; an indefinite term, which, though it distinguishes angle from length, does not serve to compare one angle with another. And just in the same manner, if it were not that the definition is more complicated, and refers to an abstract, not a visible or tangible, conception, it would immediately be seen that *ratio* is *relative* magnitude,—a term which is sufficient to distinguish the thing in question from *absolute* magnitude, but which does not give any means of comparing one thing of the kind with another. The immediate deduction of this idea is as follows: If, whenever mA lies between nB and $(n+1)$ B, it also happens that mP lies between nQ and $(n+1)$ Q, it follows that A, lying between two certain fractions of B, $\frac{n}{m}$B, and $\frac{n+1}{m}$B, then P lies between the same two fractions of Q. Or, if $m\text{A}=n\text{B}$, that is, if $\text{A}=\frac{n}{m}\text{B}$, then P is the same fraction of Q. Or we may state it thus: if B be made unity, for the measurement of A, and Q for the measurement of P, then A and P are the same numbers or fractions of their respective units.

Euclid has commenced the subject with a rough definition, as we have seen, p. 29, and the translators have spoiled it, by not distinguishing between quantity, and relative quantity; that is, by so wording the definition as to say nothing more than that ratio is a relation of magnitudes with respect to magnitude.

We now come to consider the application of the preceding notions to arithmetic. Let us first separate all that part* of arithmetic which relates to abstract and definite numbers, from the rest, and let us call it *primary* arithmetic. A little observation will shew that *abstract* number as distinguished from concrete, is really the same thing as ratio of magnitude to magnitude. What is *three*, for example? It is an idea which we obtain equally from looking at

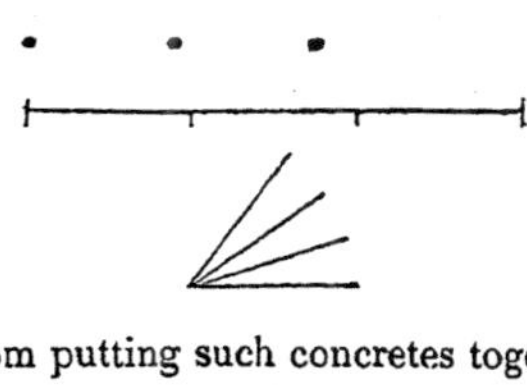

and

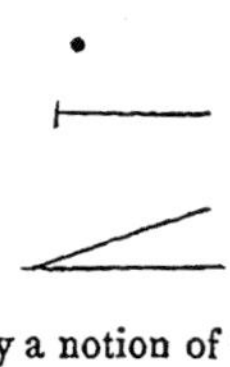

From putting such concretes together, we bring away a notion of there being the same relative magnitudes existing between the individuals

* The whole of the First Book of my *Treatise on Arithmetic*, with the exception of § 158, 165–169.

of each pair. In the first, it is *repetition,* in the second, it is length, in the third, it is opening, we are reminded of; but in all three, we say the first is *three times* the second. Now this word *times* is, in fact, a limitation, which will not do for our present purpose; it implies that we will have no other ratios except those of line to line in the series

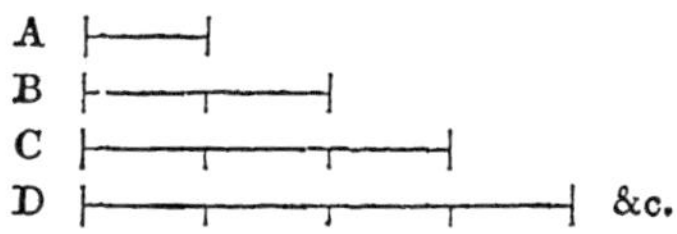

made by repetitions only: but there may be ratios which are not those of line to line in any *repetition,* how far soever carried.

Here is a point at which we are compelled to pause, to adjust the well-known terms of number to the new idea we have put upon them: *Abstract numbers are certain ratios; abstract fractions are certain other ratios: but all possible ratios are not found among numbers and fractions; whence it arises, that primary arithmetic, though it may be, so far as it goes, a theory of ratios, is not a theory of all ratios, nor are its operations such as can be performed upon all ratios.*

That ratios are magnitudes, we must have supposed from the beginning, seeing that they bear the terms equal, greater, and less. But there was still this defect, that our test of A being to B more than C to D, was one which left us with no idea *how much* more A was to B than C to D; which amounts but to this, that we could not define the *ratio* of *ratios* without having first defined *ratio.* But, in like manner as arithmetic was made the guide to that notion which is properly* called the ratio of incommensurable quantities, so will the ratio of two ratios in arithmetic lead us, after a little consideration, to the meaning of the ratio of ratios of incommensurables.

When we say *two,* we refer to the repetitions of the smaller in a ratio of magnitudes, thus visibly related:

When we say twice two, there is a change of idiom in our language. It might be, instead of *twice two* is *four, two twos* are *four;* that is, where there exists that idea of relative magnitude which we signify by

* Consistently; so as to couple with operations upon problems of commensurables those operations which apply to the same problems upon incommensurables.

two, let the idea of *relation* be coupled with the idea of a *larger relation*, in exactly the same manner as our idea of *magnitude*, when we look at ——, is increased when we look at —— ——; and we shall then, by considering the result as one of relative magnitude, be led to the idea of the relation between —— —— —— —— and ——. This, of course, does not give a better comprehension of twice two is four; but what it explains is, that we are using the term *ratio* in a consistent sense, when we say that the ratio of 2 to 1, increased in the ratio of 2 to 1, is the same as the ratio of 4 to 1; and, generally, that the ratio of m to 1, increased in the ratio of n to 1, is the ratio of mn to 1. And the notion of relative magnitude contained in the words, ratio of m to p, must be the same as that contained in the words, ratio of mn to pn; and, conversely, the notion in the latter is that implied in the former. I doubt if any thing that deserves the name of proof can be given of this proposition, which seems to be worthy the name of an axiom. What idea we form of magnitude as portion of magnitude from A and B, the same do we form from 2A and 2B. Nor can I imagine these propositions extended to fractions in any more fundamental manner than by observing, that as $\frac{m}{n}$ taken $\frac{p}{q}$ times is $\frac{mp}{nq}$ times (times mean times, or parts of times, either separately or both together,) a unit, the ratio of $\frac{m}{n}$ to 1, altered in the ratio of $\frac{p}{q}$ to 1, is the ratio of $\frac{mp}{nq}$ to 1; or that the ratio of m to n, altered in the ratio of p to q, is the ratio of mp to nq. These are propositions in which the line between deduction and mere establishment of the synonymous character of terms is very indefinite. I recommend the student to examine his own idea of what he would have meant by "the proportion of 3 to 2 increased in the proportion of 5 to 4, is the proportion of 15 to 8." If he be a metaphysician, I refer him to his oracle, on condition only that the response shall not contradict the preceding proposition.

The multiplication of m and n is, then, the alteration of the ratio of m to 1 in the proportion of n to 1; and the ratio of magnitudes mA and nA is the same as the ratio of magnitudes mB and nB, and of m to n. Hence, to alter $m\mathrm{A} : n\mathrm{A}$ (which is $m : n$) in the ratio of pB to qB, which is ($p : q$), is the formation of $mp : nq$, or mpA to nqA, or mpB to nqB. Now, this is precisely what Euclid has termed the *composition* of these ratios; for, let $mA : nA :: vB : pB$, then $vB : pB$ compounded with $pB : qB$, is $vB : qB$, or $v : q$. But

$$mA : nA :: m : n \qquad vB : pB :: v : p$$

Therefore $\quad m : n$ is $v : p$ or $\frac{m}{n} : 1$ is $\frac{v}{p} : 1$

$v = \frac{pm}{n}$ $\quad v : q$ is $\frac{pm}{n} : q$ or $pm : nq$ or $pmA : nqA$

or $\quad pmB : nqB$.

Hence, *composition* is multiplication of terms, when the ratios are those of number to number. Let, then, composition of ratios stand for multiplication of terms, and be considered as the corresponding operation in the case of incommensurable magnitudes.

Prove from this, that if $U : A$ and $U : B$ be compounded, giving $U : C$, that when $A = aU$ and $B = bU$, we have $C = abU$, and that if $U : A$ and $B : U$ be thus compounded, giving $U : D$, we have $D = \frac{a}{b}U$, $D : U :: \frac{a}{b} : 1$; in which operations, corresponding to multiplication and division.

It may be a matter of some curiosity to know whether Euclid carried with him the notion of multiplication of numbers in the composition of ratios. In the Fifth Book, the notion of the numerical magnitude of a ratio is entirely suppressed, except only in the single word πηλικοτης (see page 29.) Composition* is defined to be the taking an antecedent of one ratio with the consequent of another; and it is not even specified that the intermediate terms are to be the same. But in the Sixth Book we find composition, or collocation of ratios, to mean the *multiplication of their quantuplicities* (see page 29).

* Σύνθεσις λόγου ἐστὶ λῆψις τοῦ ἡγουμένου μετὰ τοῦ ἑπομένου ὡς ενὸς προς αὐτὸ τὸ ἑπόμενον.—V. Def. 15.

Λόγος ἐκ λόγων συγκεῖσθαι λέγεται ὅταν αἱ τῶν λόγων πηλικότητες ἐφ' ἑαυτὰς πολλαπλασιασθεῖσαι ποιῶσι τινά.—VI. Def. 5.

The second of these definitions has usually been omitted in modern editions. But it is worthy of remark, that, in the first, to compound is συντιθεσθαι; in the second, συγκεισθαι; and the second is the word afterwards used by Euclid, though in the sense of the first. The reason of the omission appears to have been a disposition on the part of commentators to consider Euclid as a perfect book, and every thing which did not accord with their notions of perfection, as the work of unskilful editors or interpolators.

The addition and subtraction of ratios can only be primarily conceived when the latter terms of the ratios are alike. Thus,

we must imagine the idea of relative magnitude given by BC compared with A, and by CD compared with A, to be put together, in order to make up the relative magnitude of BD to A. Addition and subtraction are, as to ratios, ideas not so simple as multiplication and division. Shew that the preceding is the only way in which $m : 1$, increased in the ratio of $n : 1$, will give $mn : 1$, consistently with the notion of multiplication of whole numbers being successive additions.

When ratios have not the same consequents, they must be reduced to the same consequents. Thus, A : B and C : D are added by taking A : B :: P : Z and C : D :: Q : Z, and P + Q : Z is the sum of the ratios. This answers to addition of fractions.

Let P be the mean proportional between A and B, meaning that A : P as P : B. It may be proved, as in page 60, that there must be such a magnitude as this mean proportional, and we may also prove that we can find A : P as P : Q, and P : Q as Q : B, thus forming two mean proportionals. It is readily proved, that if $A = aU$ and $B = bU$, then $P = cU$ where $cc = ab$. If, then, ab be a number or fraction which has a square root, P can be found commensurable with A and B; but if ab have no square root, number or fraction, then P is incommensurable with A and B, but not, therefore, unassignable as a magnitude, though unassignable as a numerical fraction of A or B. Consequently, when we speak of $\sqrt{2}$, it must be with reference to magnitude, and we mean $\sqrt{2}\,M$, an accurate representative (if we choose to define it so) of the mean proportional between M and 2 M. Similarly, when there are two mean proportionals, we find P, if $A = aU$ and $B = bU$, to be cU where $ccc = ab$, and this is incommensurable unless ab be a cube number or fraction. But we may define $\sqrt[3]{2}\,M$ to be the first of two mean proportionals between M and 2 M; and so on.

Are we, then, to use long processes and comparatively obscure definitions, whenever the ratios of a problem are incommensurables? By no means; we proceed to shew that it may always be made possible to let the processes of arithmetic (or rather of algebra) be used as if the ratios in question were commensurable; and that we may thus deduce a result which may either be interpreted strictly at the end of the process, or made to give a result as near as we please to the truth in arithmetical terms. Let us suppose this PROBLEM : Two pounds are spent in buying yards of stuff, and as many yards are bought as shillings are given for a yard. Let x be the number of yards, then x yards at x shillings a yard, gives xx shillings; whence $xx = 40$, which is arithmetically impossible. Now, turn from numbers of pounds to quantities of silver, and let S be the silver in a shilling, X that in the price given; let L be a yard, and Y the length bought. Then it is required that 40 S should be given, and that X should bear the same ratio to S as Y bears to L. Now, if X be given for L, what must be given for Y? Take P of such relative magnitude to X, as Y is to L; that is, let

$$\mathrm{L : Y :: X : P = 40S}$$

But as $\mathrm{L : Y :: S : X}$ Therefore $\mathrm{S : X :: X : 40S}$

or X must be a mean proportional between S and 40S. Now, if we make our symbols general, and let x stand for any ratio, numerically possible or not, but proceed as we should do if it were arithmetical, we proceed as in the first case, and find $x = \sqrt{40}$, which, being interpreted as a magnitude, with reference to its ratio to S, means, when the symbols are general, $\sqrt{40}\,\mathrm{S}$, the mean proportional between S and 40S. If we wish for an approximate numerical result, we must suppose $40 + a$ to be the sum, where $40 + a$ has a square root, and then we have $x = \sqrt{40 + a}$; and since a may be made as small as we please, we can make this problem as near the given one as we please.

The following table should be attentively considered. In the first column, an incommensurable ratio x, of X to U, is given, or a function of it and other ratios, under arithmetical symbols; in the second is the ratio which the function really gives, when the symbols on the first side are extended in meaning.

x or $x : 1$	the ratio of	X	to	U
y .. $y : 1$	....	Y	..	U
z .. $z : 1$	....	Z	..	V

$\frac{1}{x}$	$\frac{1}{x} : 1 :: 1 : x$ or $U : X$ as $\frac{1}{x} : 1$
x^2	compounded of $x : 1$ and $x : 1$ or $X : U$ and $X : U$. Let $U : X :: X : P$ then $P : X$ and $X : U$ compounded give $P : U$ or $x^2 : 1$ is the ratio which a third proportional to U and X bears to U.
xyz	$(x : 1)\ (y : 1)\ (z : 1)$ ratio compd. of $X : U$, $Y : U$, and $Z : V$. Let $X : U :: P : Y$; the above is then compounded of $P : U$ and $Z : V$. Let $P : U :: Q : Z$. The result is then $Q : V$ or $xyz : 1$ is $Q : V$
	P a fourth prop. to U, X, and Y
	Q .... U, P, .. Z
$xy + yz$	$xy : 1$ is $P : U$ when $U : X :: Y : P$
	$yz : 1$ is $Q : V$ .... $U : Y :: Z : Q$
	Take $Q : V :: M : U$ or $V : Q :: U : M$
	$P + M : U$ is the ratio required.
$\frac{x}{z}$	$x\frac{1}{z} : 1$ compd. of $X : U$ and $V : Z$
	Let $X : U :: P : V$ and $P : Z$ is the ratio required.

Now, we have assumed the operations of finding a fourth proportional, a mean proportional, two mean proportionals, &c. Whether these can be done, or whether any or all cannot be done, is a question for every particular application. In arithmetic, we will suppose the data arithmetical; a fourth proportional can always be found. In geometry, a fourth proportional can be found to lines or rectilinear areas; but not to angles, &c. And a mean proportional cannot generally be found in arithmetic, but can be found in geometry, between two straight lines, or two rectilinear areas. But two mean proportionals cannot be found in geometry or in arithmetic.

It must be remembered, that while we are here speaking of geometry or arithmetic, we are not speaking of every conception we can form of these sciences, but of the subjects as limited by the definitions of what it has been agreed shall be called arithmetic and geometry. Elementary arithmetic means the science of numbers and fractions: elementary geometry, the science of space, so far as the same has properties which can be deduced by allowing of *fixed* straight lines and circles. To say that an angle cannot be trisected *geometrically*, means, that it cannot be trisected by means of straight lines and circles as defined. But there is an abundance of curves, the stipulation to draw any one of which would secure the means of trisecting an angle. And, by simply granting that a circle should be allowed to roll along a straight line, and that the curve described by one of its points should be granted, we can either square the circle, or find the ratio of any two arcs. And, just in the same way, if we were to define a journey to be 100 miles or less, it would be perfectly true that we could not make a journey from London to York, but that we could from London to Brighton.

It is surely time that the verbal distinction between different parts of the same sciences should be done away with. Every conception which can be shewn to be not self contradictory, can be as easily realised by assumption as the drawing of a circle, which is itself a perfect geometrical idea, and can only be roughly represented by mechanical means. Whatever can be distinctly conceived, exists for all mental purposes; whatever can be approximately found, for all practical uses.

It may be worth while to make the student remark the close similarity which exists between the process in page 64, and that by which we enlarge our ideas in algebra, from the simple consideration of numerical magnitude to that of positive and negative quantities. In both, we set out with a notation insufficient to express all the results of problems; in both, this circumstance is marked by the appearance of unexplained results, the examination of which, on wider grounds, shews the necessity for attaching more extensive ideas to symbols; and in both, the partial view first taken is wholly included in the more general one: while in both, the processes conducted under the wider meanings are precisely the same in form and rules as those which are restricted to the original meanings of the symbols. The principal difference is, that in extending arithmetic to the general science of ratios, we are not engaged in interpreting difficulties arising from contradictions, but from results which are only approximately attainable. But in both the reason is, that we set out with our symbols so constructed, that we cannot undertake a problem without tacitly dictating conditions to the result. In beginning algebra, we make quantities indeterminate in magnitude, with symbols of operation so fixed in meaning, that they cannot be used without an assumption that we know which is the greater and which is the less of two unknown quantities. We have, therefore, to examine the different cases of problems which present different results according as one datum is greater or less than another; and thus we obtain those extensions of meaning which will make the problems and the symbols equally general. In beginning arithmetic, we invent no symbols of ratio, except those which represent the ratios of magnitudes formed by the repetitions of a given magnitude. These we find to be not sufficient to represent all ratios; though it is shewn that we can make them represent any ratio which magnitudes can have, as nearly as we please. The invention of new symbols of ratio must require the generalisation of operations; that is, we cannot speak of *multiplication* or *division* of ratios generally, while these words have a definition which applies only to ratios of repetitions, or commensurable ratios.

There is a difference between the *impossible* of primary arithmetic, and that of geometry. The first is unattainable by a restricted definition, the second by restricting the cases of general definitions which shall be allowed to be used. In arithmetic, we attempt a science of relative magnitudes, by running from the general notion of relative magnitude to the more precise and easy notion of the relative magnitudes of one certain set of magnitudes, A, an arbitrary, A + A, A + A + A, &c. We are very soon taught that our symbols will not express all ratios, that is, if we have a general notion of ratio to think about: whence our definitions are not sufficiently extensive. But in geometry, having assumed notions and definitions from which we cannot help conceiving an infinite number of different lines and curves, we immediately proceed to cut ourselves off from the use of all except the straight line and circle; that is, the straight line between or beyond two given points, and the circle which has a given centre and a given radial line. Until these demands or postulates are looked upon as *restrictions*, their sense is never understood. (See the Appendix.)

This difference is, however, not very essential; since it is much the same whether we define in too limited a manner, or whether we limit ourselves to the use of only a part of a general definition. We shall in the sequel discard the restrictive postulates, and suppose ourselves able to draw any line which we can shew to be made by the motion of a point.

The method by which Euclid first exhibits four proportional

straight lines, though elegant and ingenious, has not the advantage of exhibiting the notion of ratio directly applied to two straight lines. The following theorem is directly proved from the first book, and may be made the guide. If a series of parallels cut off consecutive equal parts from any one line which they cut, they do the same from every other. This premised, suppose any two lines OA, OB, and take a succession of lines equal to OA and OB, drawing through every point a parallel to a given line. Draw any other line, OCD, intersecting all the parallels: from which the preliminary proposition shews, that whatever multiple O*a* is of OA, the same is O*c* of OC; and whatever O*b* is of OB, the same is O*d* of OD. And if O*a* be greater than, equal to, or less than O*b*, O*c* is greater than, equal to, or less than, O*d*. Hence the definition of equal ratios applies precisely to the lines OA, OB, OC, and OD, which are, therefore, proportionals. This gives the construction of Book VI. Prop. 12, or one analogous to it.

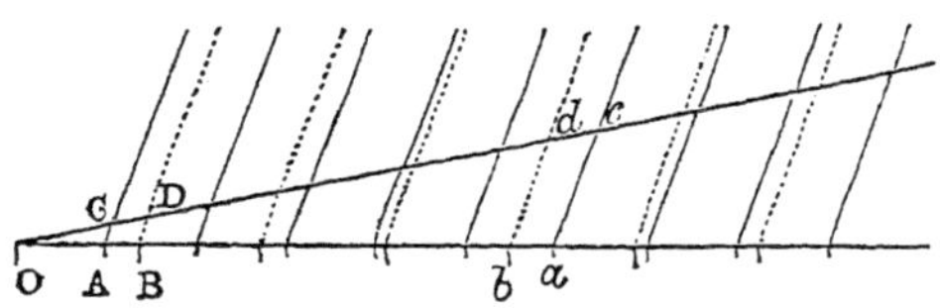

The method of finding a mean proportional between two straight lines is given in Prop. 13; but as we now wish to make the straight line the foundation of general conceptions of magnitude, we shall pass at once to those considerations which involve any number of mean proportionals. It adds considerably to the interest of this part of the subject, that we are thus brought to the notions on which the first theory of logarithms was founded.

Let there be any number of lines, V, V_1, V_2, V_3, ... in continued proportion; that is, let all the ratios of V to V_1, V_1 to V_2, V_2 to V_3, &c. be the same. And let V_1 be greater than V; in which case V_2 is greater than V_1, &c. If V_1 were equal to V, then would V_2 be equal to V_1, &c. And, first, we have the following

THEOREM. By however little V_1 exceeds V, the series V, V_1, &c. is a series of magnitudes increasing without limit: so that, however great A may be, a point may be attained from and after which every term is greater than A: but in all cases whatsoever, V_1 may be taken so near to V, that the terms of the series V, V_1, &c. between which A lies, shall be as near to A in magnitude as we please.

Firstly, the series increases without limit. For, since $V : V_1 :: V_1 : V_2$, and V and V_2 are the greatest and least, we have

$$V + V_2 \text{ is greater than } V_1 + V_1$$

or

$$V_2 - V_1 \text{ is greater than } V_1 - V$$

Or, V_2 exceeds V_1 by more than V_1 exceeds V. Similarly, V_3 exceeds V_2 by more than V_2 exceeds V_1; and so on. But if to V were added continually the same quantity, the result would come in time to exceed any given magnitude; still more when a greater quantity is added at every step.

Secondly, since then we come at last to V_n less than A, while V_{n+1} exceeds A, it is plain that A will not differ from either by so much as they differ from each other. But because

$$V_n : V_{n+1} :: V : V_1$$

we have

$$V_{n+1} - V_n : V_n :: V_1 - V : V$$

If then $V_1 - V$ be so small that $m(V_1 - V)$ shall not exceed V, neither will $m(V_{n+1} - V_n)$ exceed V_n, and of course not A. Let m be any given number, however great, and let $V_1 - V$ be less than the mth part of V; then will $V_{n+1} - V_n$ be less than the mth part of A; or, by taking m sufficiently great, may be made as small as we please. Whence the second part of the theorem.

THEOREM. In the preceding series, the selection

$$V \quad V_n \quad V_{2n} \quad V_{3n} \quad \&c.$$

constitutes a similar series of continued proportionals. For, since any two consecutives in the upper line next given are proportiona to those under them in the lower,

$$\begin{matrix} V, & V_1, & V_2 & \cdots\cdots & V_n \\ V_n & V_{n+1} & V_{n+2} & \cdots\cdots & V_{2n} \end{matrix}$$

we have (XXII.) $V : V_n :: V_n : V_{2n}$: and so on.

If between each of the terms of the series we insert the same number of mean proportionals, the series thus formed will have the same properties as the original. Let us say we insert two mean proportionals between each two terms. Then we have

$$V \ K \ K' \ V_1 \ L \ L' \ V_2 \ M \ M' \ V_3 \ \ldots\ldots$$

Now the only question about the continuance of the same ratio from term to term is in the ratios $V_1 : L$, $V_2 : M$, &c. But I say that since

$$\begin{matrix} V : K :: K : K' :: K' : V_1 \\ V_1 : L :: L : L' :: L' : V_2 \end{matrix} \quad \text{and} \quad \begin{matrix} V : V_1 \\ V_1 : V_2 \end{matrix}$$

that $V : K :: V_1 : L$. For if not, let these latter ratios differ; say V is to K more than V_1 is to L. Then is K to K′ more than L is to L′; and hence (presently will be shewn) the ratio compounded of V to K and K to K′, or $V : K'$, is greater than that compounded of $V_1 : L$ and $L : L'$ or $V : L'$. Similarly, V to K′ and K′ to V_1 being more than $V_1 : L'$ and $L' : V_2$, we have $V : V_1$ is more than V_1 to V_2, which is not true. Therefore V is not to K more than V_1 to L; a similar process shews that it is not less: consequently,

$$V : K :: V_1 : L$$

or the continuance of the primary ratio is uninterrupted.

The theorem assumed in the above is thus proved. If $A : B$ more than $P : Q$ we have mA greater than nB, while mP is less than nQ; or any other descending assertion. And if $B : C$ more than $Q : R$, we have xB greater than yC, while xQ is less than yR. Or we have

$$\begin{matrix} mxA \text{ greater than } nxB, & nxB \text{ greater than } nyC, & \text{or } mxA \text{ greater than } nyC \\ mxP \text{ less than } nxQ, & nxQ \text{ less than } nyR, & \text{or } mxP \text{ less than } nyR \end{matrix}$$

that is, A is to C more than P is to R; which is what we assumed.

If then we insert a mean proportional between V and A, giving

$$V \quad M \quad A$$

if between each we insert a mean proportional, we have

$$V \quad M' \quad M \quad M'' \quad A$$

If we proceed in this way, we shall come at last to a series of the form

$$V \quad V_1 \quad V_2 \quad \ldots\ldots\ldots \quad V_{n-1} \ (V_n = A)$$

in which no two quantities differ by so much as a given quantity K. We can actually insert one mean proportional between any two quantities; it is done in geometry between two lines, and (page 60) two magnitudes of any sort may be made (one being given) proportional to two lines. Thus, let A, B, C, be continually proportional lines, or let B be a mean proportional between A and C. Then if A and C were taken proportional to (say angles) M and K, it follows that if A : B : : M : L, that M, L, and K are continued proportionals, by a proof of the sort given in the lemma of the last theorem. Granting, then, that every two magnitudes have one mean proportional, we may now shew that they have any number of intermediate proportionals; as follows:

We set out with 2 quantities, and the first insertion adds 1, the second 2, the third 2^2, the fourth 2^3 .... and the nth 2^{n-1}. Consequently, n complete insertions add

$$1+2+2^2+ \cdots\cdots +2^{n-1} \quad \text{or} \quad 2^n-1$$

to the first 2; giving 2^n+1 in all. Now, let us suppose that 2^n+1 divided by p leaves a quotient q, and a remainder r which is not greater than p. Consequently, we have for the whole number (V and A inclusive) after n insertions,

$$v=pq+r \quad \text{which is also} \quad p(q+1)-(p-r)$$

and $p-r$ is also not greater than p; and V_m=A when $m=v$ and is greater or less than A, according as m is greater or less than v. If then out of the series (the proportion being continued up to $V_{p(q+1)}$) we select

$$\begin{array}{llll} V & V_q & V_{2q} \quad \cdots\cdots\cdots\cdots & (V_{pq} \quad \text{less than} \quad A) \\ V & V_{q+1} & V_{2(q+1)} \quad \cdots\cdots\cdots & (V_{p(q+1)} \text{ greater than } A) \end{array}$$

We see V and V_{pq}, and V and $V_{p(q+1)}$ each with $p-1$ mean proportionals inserted between them, namely,

$$V_q \quad V_{2q} \quad \cdots\cdots \quad V_{(p-1)q} \quad \text{and} \quad V_{q+1} \quad V_{2(q+1)} \quad \cdots\cdots \quad V_{(p-1)(q+1)}$$

But from V_{pq} to $V_{p(q+1)}$ there are p passages from term to term of the complete series, consequently, since each passage may be made by an augmentation less than K, the difference between the two may be made less than pK, which call Z. Hence we have the following

Theorem. To find two magnitudes, one greater and the other less than A, but differing from it by less than a given quantity Z, between each of which and V, $p-1$ mean proportionals shall exist, obtained by continual insertion of one mean proportional, continue the insertion until no two successive terms shall differ by so much as the

pth part of the quantity Z : then the quantities required and the mean proportionals shall be in the set so found.

Hence it can be shewn that there *are* $p-1$ magnitudes (whether attainable or not with any given means is not the question) which are mean proportionals between V and A. Let P_p and Q_p be magnitudes, one greater and one less than A, which have such mean proportionals, namely, let the following be continued proportionals,

$$V \quad P_1 \quad P_2 \cdots\cdots P_{p-1} \ (P_p \text{ greater than } A)$$
$$V \quad Q_1 \quad Q_2 \cdots\cdots Q_{p-1} \ (Q_p \text{ less than } A)$$

obtained by the preceding method, from which it is apparent that P_1 is greater than Q_1. Now, exactly as in page 60, if we assume X_1 to set out in value $= Q_1$, so that $V : X_p$ more than $V : A$ (X_p being the pth of the set of continued proportionals $V, X_1, X_2, \ldots\ldots$) and to change through all possible intermediate magnitudes up to $X_1 = P_1$, or $V : X_p$ less than $V : A$, there is but this alternative; EITHER at some intermediate point $V : X_p$ as $V : A$, or $X_p = A$, OR, there is a point at which $V : X_n$ more than $V : A$, being always less when X_1 is greater by any magnitude however small. The latter may be disproved, or the former proved, as in the page cited.

To resume the original subject. It appears, then, 1st, that if between V and A we continually insert mean proportionals, in such manner that at every step one mean proportional is inserted between every two consecutive results of the preceding step. 2d, If the series be continued beyond A, preserving still the same ratio between the consecutive terms of the continuation which exists between consecutive terms lying between V and A; then will this process leave us at last with a series of consecutive proportionals, having consecutive terms so near together in magnitude, that every magnitude lying between V and any we please to name, shall have a term of the series differing from it by less than Z, however small Z may be.

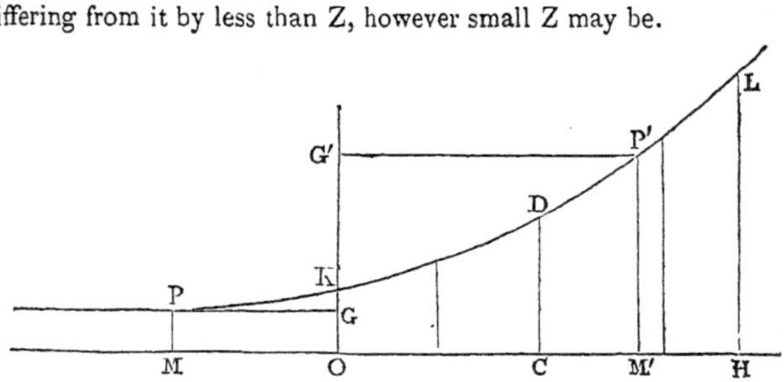

Let us now make O K and H L perpendicular to any chosen line O M, and let V be the line O K, A the line H L. Bisect O M in C, and erect C D the mean proportional between O K and H L. Bisect O C and C H, and erect the mean proportionals between O K and C D, and between C D and H L. Continue this process, and we shall thus get an increasing number of points between K and L, which will soon give to the eye the idea of a curve line rising from K to L. When we have thus divided O H into 2^n parts, by n insertions, giving $2^n + 1$ lines, we may, by setting off portions equal to those intercepted in O H, continue that line on one side and the other, and thus continue the scale of proportionals and the series of points on one side and on the other of O and H. However far we may go we can never complete this curve; but if we admit that *a curve exists*, wherever a series of points can be laid down, as many as we please, and consecutively as near as we please, then we have a right to assume this curve as existing, and, *for purposes of reasoning*, as constructed. Call this the *exponential* curve, (*exponere*, to set forth), which *expounds* ratios, a phrase to which we shall presently give meaning. That the student may not suppose we are using an old word in a new sense, it is necessary to inform him that this curve, or rather the process which we have illustrated by it, is older than the algebraical symbol a^x, and that x gets the name of exponent from it. We shall presently see the analogy.

The exponential curve being given, every line O G has its place M P among the *ordinates* of the curve, and its *abscissa* O M, which expounds or sets forth that place. From the nature of the formation, it is evident that a given line has but one exponent, and that the order of magnitude of lines (to the right of O), is also that of their exponents.

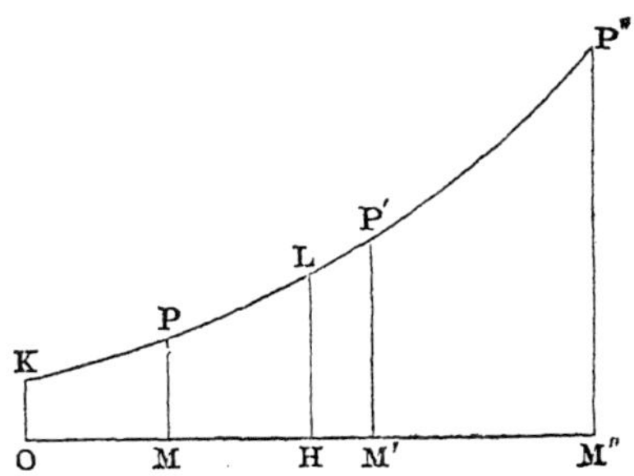

And the main property of the curve is this: that a fourth proportional to any three lines (V being one), OK, MP, M′P′, may be found by adding the exponents OM and OM′ (making OM″), and finding the line M″P″ expounded by that sum. To prove this, make n sets of insertions in OH, and suppose MP to lie between V_m and V_{m+1}, while M′P′ lies between $V_{m'}$ and $V_{m'+1}$. Now, in the series of continued proportionals,

$$V \; V_1 \; V_2 \;\; \cdots\cdots\cdots \;\; (V_{2^n+1} = A) \; V_{2^n+2} \;\; \cdots$$

I say that $V : V_m : V_{m'} : V_{m+m'}$

For $V \quad V_1 \quad V_2 \quad \cdots\cdots \quad V_{m-1} \quad V_m$

$V_{m'} \quad V_{m'+1} \; V_{m'+2} \quad \cdots\cdots \quad V_{m'+m-1} \; V_{m'+m}$

we have $V : V_1 :: V_{m'} : V_{m'+1}$ &c. &c.

whence $V : V_m : V_{m'} : V_{m+m'}$

Similarly, $V : V_{m+1} : V_{m'+1} : V_{m+m'+2}$

Now, by a lemma we shall presently shew, since MP lies between V_m and V_{m+1}, and M′P′ lies between $V_{m'}$ and $V_{m'+1}$, the fourth proportional required lies between $V_{m+m'}$ and $V_{m+m'+2}$. Let K be the value of one of the last subdivisions of OH; then we have supposed OM to lie between mK and $(m+1)$K, and OM′ between m'K and $(m'+1)$K. The preceding makes it evident that the fourth proportional has an exponent between $(m+m')$K and $(m+m'+2)$K; while the sum of the exponents OM and OM′ also lies between $(m+m')$K and $(m+m'+2)$K. Since K can be made as small as we please, it must follow that the sum of the exponents is the exponent of the fourth proportional; for two different magnitudes cannot lie between two quantities which can be made as near as we please, as can $(m+m')$K and $(m+m')$K + 2K. If the two approximating magnitudes approach to each other, keeping one of two *different* magnitudes between them, they must, at last, leave out the other.

The lemma alluded to is as follows: If

$$A : B :: C : D$$

and $\quad A : B + B' :: C + C' : D + D'$

Then if A, B + X, C + Y, D + Z, be also proportionals, where X and Y are less than B′ and C′, then Z must be less than D′; for A is to B + X more than A is to B + B′; or (substituting equal ratios), C + Y is to D + Z more than C + C′ is to D + D′. Still more is C + C′ (remember that C′ is greater than Y) to D + Z more than C + C′ to D + D′; that is, D + Z is less than D + D′, or Z less than D′.

The following property we leave to the student to deduce from the last. If there be any three lines, X_1 X_2 X_3, expounding Y_1 Y_2 Y_3, any lines whatsoever greater than V, then the exponent of the fourth proportional is $X_2 + X_3 - X_1$.

These are all properties of algebraical exponents, or of logarithms, (λογων αριθμοι, numbers expounding ratios). We shall now make it appear, that the line expounded by x is of the form a^x.

Let the numerical symbol of V or OK be v; let that of HL or A be a. Then, if arithmetical mean proportions be continually inserted, we have

$$v \qquad (av)^{\frac{1}{2}} \qquad a$$

$$v \qquad a^{\frac{1}{4}}v^{\frac{3}{4}} \quad a^{\frac{2}{4}}v^{\frac{2}{4}} \quad a^{\frac{3}{4}}v^{\frac{1}{4}} \qquad a$$

$$v \qquad a^{\frac{1}{8}}v^{\frac{1}{8}} \quad a^{\frac{2}{8}}v^{\frac{6}{8}} \quad a^{\frac{3}{8}}v^{\frac{5}{8}} \quad \ldots\ldots\ldots \quad a^{\frac{7}{8}}v^{\frac{1}{8}} \qquad a$$

or generally, when $2^n - 1$ (say $p - 1$) mean proportionals are inserted between v and a, the mth of these proportionals is

$$(2^n = p) \qquad a^{\frac{m}{p}}v^{1-\frac{m}{p}} \quad \text{which is} \quad vk^{\frac{m}{p}}$$

if we suppose $a = vk$. Now, let us suppose a number y thus expounded by x; and after n insertions, let this number x lie between $m\alpha$ and $(m+1)\alpha$, α being the pth part of OH, (let OH be c). We have then

$$x \quad \text{lies between} \quad m\frac{c}{p} \quad \text{and} \quad (m+1)\frac{c}{p}$$

$$\text{or} \quad \text{between} \quad c\frac{m}{p} \quad \text{and} \quad c\frac{m}{p} + \frac{c}{p}$$

or
$$x = c\frac{m}{p} + \beta \qquad \left(\beta < \frac{c}{p}\right)$$

Therefore, $\dfrac{m}{p} = \dfrac{x}{c} - \dfrac{\beta}{c}$

Consequently the number expounded by $m\alpha$, or $m\frac{c}{p}$, or $x - \beta$ is

$$v\,k^{\frac{x-\beta}{c}}$$

and since β diminishes without limit as the insertions continue, the number expounded by x is $vk^{\frac{x}{c}}$. That is, if we adopt general numerical symbols, let OM $= x$, MP $= y$, and we have

$$y = v\,k^{\frac{x}{c}}$$

or, if we let OK represent the linear unit ($v = 1$), and let OH $=$ OK $= 1$, HL $= 10$, or $k = 10$, we have

$$y = 10^{x}$$

or x is the common logarithm of y.

From the curve we see how it is that magnitudes less than V are expounded by negative quantities, with other well-known properties of logarithms.

We see then, that the assertion "the common logarithm of 2 is ·30103 very nearly," may be thus made; which is perhaps the most distinct view that can be given of a *numerical* logarithm. If we make 10V the hundred thousandth magnitude in a series of proportionals,

$$V,\ V_1,\ V_2,\ \ldots\ldots\ (V_{100,000} = 10)\ V_{100,001},\ \&c.$$

then will the 30103rd of these proportionals, or V_{30103}, be very nearly equal to 2 V.

If we chose, we might, granting that the exponential curve can be constructed, make VkX by definition the line MP; where X stands for OM, and k for the ratio of HL to OK. From this it would readily be deduced, that when k represents a commensurable ratio, and X is $\frac{m}{p}$ linear units where $\sqrt[p]{k^m}$ has an arithmetical existence, the results of this theory are the same as those of common algebra. And from hence it appears, that the science known by the name of the application of algebra to geometry (of which it is the foundation, that a linear unit being given, every expression of algebra may be considered as a length, or at least the symbol of the ratio of a length to that unit) does, in point of fact, make this additional assumption, while an application of geometry (with this assumption) to algebra, would take away all want of rigorous conception of the meaning of algebraical formulæ, so far as the meaning of the exponent is concerned.

The view above given is very nearly that by which logarithms were first calculated, but the method was not so general. The *natural* logarithms (see my *Algebra*, p. 226) arose thus. If we suppose a very large number of mean proportionals, then V and V_1 will be very nearly equal. Let $V_1 = V + X$, then if we assume O H, so that X shall expound $V + X$ when X is very small, or more correctly, if we suppose the limit of $(V+X)$ divided by the magnitude expounded by X, as X diminishes without limit, to be unity, we have the first, or Napier's system.

APPENDIX.

ON THE DEFINITIONS, POSTULATES, AND AXIOMS OF EUCLID.

I HERE propose to endeavour to make such a subdivision of the definitions, &c. at the beginning of the First Book, as may enable the student to review the reasoning of the whole.

I shall consider the 10th and 11th axioms as among the postulates, firstly, because some old manuscripts support this change; secondly, because the older translations (from the Arabic) support it also, and even place the 12th axiom in the same list; thirdly, because it is utterly impossible to place them in Euclid's list of *common notions*. For he uses no such word as *axiom* (Greek though it be), but calls "the whole is greater than its part," κοινη ἐννοια, that which is in the conceptions of every one. Now, what is the probability that he considered "all right angles are equal," as a truth familiar to the understanding of every beginner in geometry? His *postulates* (αιτηματα, demands) do, according to the etymology of the word, include those axioms, if not the 12th also.

I also place out of view the axioms which belong to all kinds of magnitude as much as to space, namely, from the 1st to the 9th inclusive. There remains then in the shape of limitation, or assumption, six postulates, namely, three which I will call *restrictive*, being those commonly called *postulates*,* and three *assumptions*, being the 10th, 11th, and 12th axioms, so called.

Some of the definitions contain assumptions of certain conceptions existing to which names are to be given; namely, those of a point, a

* I have seen the word postulate defined as a self-evident problem; and axiom as a self-evident theorem. This definition is derived from the character of the postulates and axioms as usually given; but from no other source.

line, the extremities of a line, a straight line, a surface, the extremities of a surface, a plane surface, a plane angle, a plane rectilineal angle. Others assume the possibility of certain relations existing, as will appear from the form in which they are put. I shall now give the definitions, classified with the corresponding postulates, in the manner which appears to me to be most systematic, and placing in [ ] such additions as seem requisite.

1. A *point;* an indefinable notion; but two persons, whatever their idea of it may be, can reason together in geometry who deny a point all parts or magnitude. *Let it be granted* that a point has no parts or magnitude, and that we are concerned with no other property of it, if there be any.

2. A *line;* also indefinable, but those whose ideas of it allow it length, and deny it breadth, can proceed. Let it be granted that all reasoning upon lines is to be founded only upon the assumption that they have length without breadth. [Thickness should have been added, but breadth may mean breadth in any direction.]

3. The extremities of a line are points. [If this define any term, it must be the term *extremities*, for the other two have been defined. To me it appears something like a theorem, as follows: That which ends a line cannot have length, for it would be a part of the line; it cannot have breadth or thickness, which a line has not; it has therefore the only qualities of a point on which we reason, or comes within the definition of a point.]

4. A *straight line;* an indefinable notion, except by the rough idea that it does not go on one side or the other of the two points, [which is no definition, because it assumes the thing in question.] Let it be granted, as a common notion, that two straight lines do not enclose a space, or have not two points in common, without having all intermediate points in common. Whatever the idea of a straight line may be, this is the only property which will be appealed to.

5. *Surface;* an indefinable notion; those whose ideas give it length and breadth, but deny it thickness, have the means of reasoning upon it in geometry.

6. The extremities of a surface are lines. (See remarks on 3.)

7. A *plane* surface; an obvious notion, roughly defined by lying evenly between bounding straight lines. [This notion, however obvious, does admit of a stricter definition. It is a surface of such kind that *any* two points in it being joined by a straight line, all

intermediate points of the straight line are on the surface. This property is tacitly appealed to throughout.]

8. A plane *angle;* the inclination, or bending towards each other of two lines in a plane. [This definition is superseded by the next; no angle except one made by straight lines is ever used.]

9. A rectilineal angle (plane), the inclination of two straight lines. [An obvious notion of *opening;* it is tacitly assumed that we know how to determine when two angles are equal, or when one of them exceeds the other, as in the fourth proposition.]

10. Right angles are those made by a straight line, called a perpendicular, which falls on another straight line, making equal angles on both sides. *Postulate;* let it be granted that all right angles are equal. [This is far from an obvious postulate; the reason for it seems to have been as follows: That two straight lines which coincide in two points coincide in every point *between them,* has been admitted; it is sufficiently obvious to sense that they coincide beyond or on each side of the two common points; that is, they coincide altogether, throughout all possible length. This seems an infinite assumption; and if it be assumed instead that all right angles are equal, it may be proved afterwards that no two straight lines have a common segment; that is, that two straight lines which coincide for any length, never afterwards separate. But it may be shewn, that the assumption of all right angles being equal, amounts to the same infinity of assumption; as follows: The right angle is by definition the half of the opening which two straight lines make, when one is the continuation of the other, as AB, BC. To assume that all right angles are equal, is to assume that the doubles

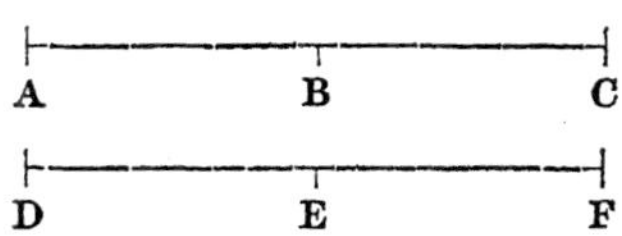

of right angles are equal; that is, that if we lay B on E, with ED coinciding with BA, then EF and BC will coincide. Now it is precisely the same thing to assume, that when AB is made to coincide with DE up to the point E, that the two coincide beyond it.

I should recommend the student to make to the assumption that two straight lines cannot coincide *in* two points without coinciding between them, the addition that they also must coincide beyond them.

It may then be directly proved that all doubles of right angles are equal, and thence that all right angles are equal.]

The definitions 11, 12, 13, 14, need no remark, being purely nominal.

14. The *circle*, a *plane* figure, having all points of its boundary (15. the *circumference*) equally distant from a given point (16. the centre) within it. [Here is tacitly a postulate, namely, that this point lies within the figure. It is also assumed in the first proposition, that if any point of a circle be within another, the two circles must intersect. There are several assumptions of this kind, which shew that Euclid did not affect that extreme *form* of accuracy which subsequent commentators have attributed to him. The assumption of a circle assumes the existence of an isosceles triangle.]

17. A *diameter* of a circle is a line passing through the centre, and terminated both ways by the circumference; it divides the circle into two equal parts, or (18. semicircles). [Here is a demonstrable theorem positively assumed. The application of one part of the circle to the other (as by revolution of one-half round the diameter) as in the fourth proposition, would prove it.]

From (19.) to (23.), the definitions are merely nominal.

24. If there be a triangle having three equal sides, let it be called *equilateral*. [In this form I give all definitions, the existence of the objects of which is to be established.]

25. An isosceles triangle is one having two sides equal.

26. A scalene triangle has the three sides unequal.] This definition is never used.]

(27.) and (28.) are nominal; (29.) tacitly refers to the thirty-second proposition; and from (30.) to (33.), should be written in the manner of (24.)

(35.) If there be two right lines, which being produced ever so far on the same side never meet, let them be called parallels. And let it be granted, that if two right lines falling upon a third make interior angles together less than two right angles, they are not parallels. [This bone of contention, when reduced to the form in which it is most palpable to the senses, is as follows: Let it be granted that two right lines which meet in a point, are not *both* parallel to any third line. This assumed, Euclid's axiom follows. For he is able to shew that the one parallel which he afterwards draws, through a point to a given line, has the property of making the two internal angles equal to right angles: *there is but one parallel;* consequently all lines which have not that property are not parallels.]

It remains to add what I have called the *restrictive* postulates. I cannot believe that Euclid, who appears to assume *very* obvious propositions, even when he might prove them, could have intended to require formally the admissions that a straight line may join two points, and may be continued, and that a circle may be drawn with a given centre and radius. If this had been the case, why not assume (Prop. IV.) that two straight lines may be drawn making equal angles with two other straight lines,—a conception more difficult than that a straight line may be drawn. I conceive, therefore, that the meaning of the three assertions commonly called postulates, is as follows: Let it be considered as intended, that no assumption of processes shall be made, except only the drawing of a straight line between two given points, the continuation of any terminated straight line to any indefinite (not given) distance, and the construction of a circle with a given centre and radius.

THE END.

LONDON :—PRINTED BY JAMES MOYES,
Castle Street, Leicester Square.

Made in United States
North Haven, CT
24 September 2026

19102539R00052